EXTREME MEASURES

D1300359

EXTREME

The Ecological Energetics of Birds and Mammals

MEASURES

BRIAN K. MCNAB

The University of Chicago Press Chicago and London

BRIAN K. MCNAB is professor emeritus in the Department of Biology at the University of
Florida. He is the author of *The Physiological Ecology of Vertebrates: A View from Energetics.*

The University of Chicago Press, Chicago 60637
The University of Chicago Press, Ltd., London
© 2012 by The University of Chicago
All rights reserved. Published 2012.
Printed in the United States of America

21 20 19 18 17 16 15 14 13 12   1 2 3 4 5

ISBN-13: 978-0-226-56122-6 (cloth)
ISBN-13: 978-0-226-56123-3 (paper)
ISBN-10: 0-226-56122-4 (cloth)
ISBN-10: 0-226-56123-2 (paper)

Library of Congress Cataloging-in-Publication Data

McNab, Brian Keith, 1932–
    Extreme measures : the ecological energetics of birds and mammals / Brian K. McNab.
        p. cm.
    Includes bibliographical references and index.
    ISBN-13: 978-0-226-56122-6 (hardcover : alkaline paper)
    ISBN-13: 978-0-226-56123-3 (paperback : alkaline paper)
    ISBN-10: 0-226-56122-4 (hardcover : alkaline paper)
    ISBN-10: 0-226-56123-2 (paperback : alkaline paper) 1. Warm-blooded animals—
Ecology. 2. Warm-blooded animals—Evolution. 3. Bioenergetics. 4. Body temperature—
Regulation. 5. Basal metabolism. I. Title.
    QP135.M38 2012
    599—dc23

                                                                            2011038631

♾ This paper meets the requirements of ANSI/NISO Z39.48-1992 (Permanence of Paper).

I dedicate this book to my colleagues at home who have encouraged me and to the many people over the years in Australia, Brazil, Brunei, Chile, Ecuador, Kenya, Indonesia, Mexico, New Zealand, Panama, Papua New Guinea, Peru, the United States, and Venezuela who have aided me, without whom little could have been accomplished. Therefore, this book is a collaborative project. However, the errors of analysis and interpretation, which many may find, are mine alone.

THE POETRY OF BIOLOGY RESIDES HIDDEN IN
OPPOSING TENSIONS, AND THE OFTEN ARDUOUS
FUN COMES FROM TRYING TO REVEAL IT.

Bernd Heinrich, *Mind of the Raven*

DON'T BE TRAPPED BY DRAMA.

Steve Jobs

CONTENTS

PREFACE

The intent of this book is to describe how the availability of energy affects the lives of, and forces restrictions on, the most energy-intensive organisms on earth: birds and mammals. Although energetics may be explored from a mechanistic or biochemical approach, my interests are principally ecological and evolutionary; that is, energetics "up," rather than "down." A concentration on the quantitative aspects of energetics has the power of integrating the degree to which organisms adjust to the circumstances that they face in the environment.

The intensity of the energetics of birds and mammals derives from their commitment to maintaining a constant body temperature by the internal generation of heat, which requires high rates of energy acquisition. However, these vertebrates must often modify their energy expenditure in response to the physical conditions they encounter, variation in the quality and quantity of available foods, the presence or absence of competitors, and the level of predation—conditions that are external to the organisms themselves. Limits on energy expenditure are also internally derived from the efficiency with which energy is used and from competing body demands, including the costs of activity, growth, and reproduction. These limits most commonly occur at very small and large body masses, in the hottest and coldest climates, or in very wet or dry environments, because all extreme conditions require an incremental expenditure of energy. Many of the adjustments organisms make require *extreme measures*, hence the title of this book. Sometimes historical commitments may limit responses to the environment, but they may also open new opportunities, such as permitting the permanent or seasonal occupation of polar and temperate environments, or dominance of terrestrial communities on oceanic islands. Energetics is especially suited to examine the probable consequences of global warming for the distribution and survival of endotherms because of its quantitative sensitivity to conditions in the environment.

In this book I will give an overall view of the energetics of birds and mammals and will attempt to integrate it with as much of their natural history as I can. I will also examine alternative views of their energetics, but in the final analysis this book will be my analysis, which will reflect the various views that I have expressed over sixty years of work in this field, subject to the changes that have been made as new information and ideas have appeared. Most of my work occurred in three areas: (1) measuring energy expenditure in more than 300 species, with special emphasis on tropical endemics, (2) suggesting quantitative analyses of the available data, and (3) attempting to place these and other measurements in an ecological context.

Throughout this book, many aspects of the energetics and life history of species will be compared with basal rate of metabolism and field energy expenditures. The data come from the most recent compilations of basal rates of metabolism in mammals (McNab 2008a) and in birds (McNab 2009a). The principal source of data on the field energy expenditures of mammals and birds is Nagy et al. (1999).

Two preliminary reviews of this book were made, one by an anonymous individual and the other by Douglas Glazier, both of whom I thank for their detailed suggestions, which led to a radical revision. I thank the many people who have helped me through the years, far too many to list individually, but especially Frank Bonaccorso, who first got me to go to Papua New Guinea, to which I returned eleven times, and with whom I studied the energetics and ecology of bats and birds. I also thank the many anonymous reviewers, both supportive and critical, of my original articles because both kinds have made a significant contribution to their improvement, although the articles' weaknesses remain my fault.

Finally, I have added a topic that never appears in a technical book: at relevant places in the text, I have placed a box to describe some of the adventures that occur when working with exotic animals in exotic places, or to express an opinion. These are the experiences that make the study of comparative biology a thrill, and I hope that their presence will not be inappropriate.

CHAPTER ONE. *Basic Energetics*

The purpose of this chapter is to provide a technical framework for the thermal biology of organisms so that a description of the responses of birds and mammals to the environments in which they live can be readily understood. Most animals belong to one of two thermal groups: they either have a body temperature that is similar to, and conforms to variations in, the dominant temperatures in the environment, or they maintain a rather high body temperature that is relatively independent of environmental temperatures (although some intermediate states exist, especially at large body masses). The vast majority of animals belong to the first group, the *poikilotherms* (*poikilos*, "variegated" or "variable"; *thermos*, "heat" or "temperature"). These animals are often called "cold-blooded" because their body temperatures usually reflect the cool ambient temperatures in which they live, and they are therefore cool to the human touch, given our core body temperature of 37°C. The second group constitutes the *homeotherms* ("*homoio*," "similar" or "constant"). These animals are often referred to as "warm-blooded" because their body temperatures are above many ambient temperatures and they therefore feel warm to our slightly cool fingertips.

Poikilotherms/ectotherms

Because the body temperatures of poikilotherms conform to a predominant environmental temperature (most environments have a variety of characteristic temperatures, but for simplicity, we will consider a local air or water temperature), the rates of chemical reactions that occur in their bodies, which are collectively referred to as the *rate of metabolism*, increase and decrease with environmental and body temperatures (fig. 1.1). Thus, the rate of metabolism in poikilotherms (measured most commonly by oxygen consumption but also by carbon dioxide production or, potentially, by heat production; see box 1.1), is high at high environmental and body temperatures and low at low temperatures (see fig. 1.1). The heat content of

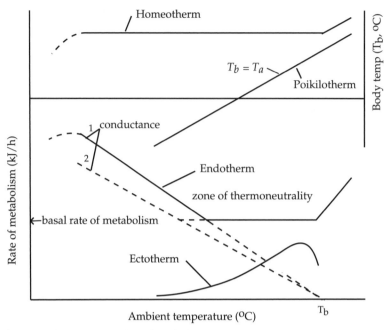

Figure 1.1. Body temperature and rate of metabolism in a poikilotherm/ectotherm and a homeotherm/ endotherm as a function of ambient temperature.

poikilotherms, then, is dictated principally by environmental temperature, which leads them to be called *ectotherms* ("outside heat"). Many complications occur in the thermal biology of ectotherms, often associated with behavior, such as the ability of some lizards to maintain a rather constant body temperature during the day by selectively absorbing solar radiation and selecting appropriate microenvironments (both of which reemphasize the ectothermic nature of their thermal biology). Furthermore, ectotherms adjust their rate of metabolism to the ambient temperatures they encounter over long periods: extended exposure to cold temperatures (cold acclimatization) tends to increase the rate of metabolism at a particular ambient temperature, although body temperature remains unchanged, whereas warm acclimatization decreases the rate of metabolism at a particular body temperature. Consequently, ectotherms in a cool, but not cold, environment tend to be about as active as ectotherms in a warm, but not hot, environment, even though their body temperatures may be quite different.

The thermal characteristics of ectotherms are also influenced by body mass. Body temperature in small species closely follows changes in ambient temperature, but as mass increases, mass-independent rates of heating and cooling (i.e., rates expressed as a percentage of values taken from standard curves) slow because of a reduced surface-to-volume ratio and increased body heat capacity. As a result, the body temperatures of large ectotherms track changes in ambient temperature more slowly than those of small species. Large ectotherms thereby gain some independence from ambient temperatures, at least in the short term. This independence is greater in aerial environments than in aquatic ones because thermal conditions in air are more complex than those in water, and because water has much greater heat capacity and conductivity.

The thermal independence of large ectotherms may approach a homeothermic condition, but one based on the thermal inertia of a large mass and a reduction in the relative rate of heat exchange with the environment. Such thermal constancy may have occurred in the largest dinosaurs, even at mass-independent rates of energy expenditure similar to those in lizards (McNab 2009c). This thermal constancy has been called "inertial homeothermy" (McNab & Auffenberg 1976) and "gigantothermy" (Paladino et al. 1990). Although this book principally concerns the thermal behavior and energetics of the homeothermic birds and mammals, the potential role of inertial homeothermy in the evolution of mammalian (and avian?) thermal behaviors will reappear in chapter 15.

Homeotherms/endotherms

Homeotherms actively maintain a rather constant body temperature by adjusting their rates of heat production and loss (see fig. 1.1). As a consequence, these vertebrates are called *endotherms* ("inside heat"). Thus, whereas some ectotherms maintain a rate of metabolism somewhat independent of environmental temperatures (through acclimatization), while their body temperatures vary, endotherms maintain a constant body temperature by varying their rate of metabolism, which makes endothermy a much more energy-demanding behavior than ectothermy. Endothermy and its associated level of activity are the principal bases for the energy intensity of birds and mammals. It clearly permits endothermic vertebrates to have an active life in harsh temperate and polar climates, but only if they can afford its cost. In persistently warm environments, ectotherms may be the ecological equal of, or even replace, endotherms, a situation most common on oceanic islands because most continental endotherms have difficulty reaching oceanic islands (see chap. 9).

The energetics of birds and mammals has been often described. For body temperature to remain independent of ambient temperature, the rate of heat production must balance the rate of heat loss. Because the rate of heat loss is proportional to the temperature differential between the body and the environment ($\Delta T = T_b - T_a$), the rate of metabolism must be proportional to ΔT for body temperature to remain constant (fig. 1.2A). Endotherms

BOX 1.1. ···

How to Measure Rates of Metabolism

In theory, rates of metabolism can be measured in a variety of ways. These methods include direct measurements of heat production as well as indirect methods such as the production of CO_2 and the consumption of O_2. The most accurate indirect method requires the measurement of both CO_2 and O_2 because the caloric or joulic equivalency of gas exchange depends on the respiratory coefficient, the ratio of CO_2 to O_2, which depends in turn on the chemical makeup of the food being metabolized. However, most measurements made are of oxygen consumption alone, which is approximately equal to 20 joules/mL O_2.

Oxygen consumption is usually measured by placing an animal in a chamber, the air of which is sucked out by a pump and replaced by room air. This system is called a "open" system because the chamber is open to the atmosphere. If the chamber is closed to the atmosphere, oxygen is metered into the system from a cylinder; this system is called a "closed" system and requires that the chamber have no leaks. The air coming out of the chamber is scrubbed of CO_2 and then of water vapor, its flow rate is measured, and it is sent to an oxygen analyzer, which measures the amount of oxygen it contains. The rate of metabolism is proportional to the difference in oxygen content between room air and the air coming out of the chamber, and this difference is multiplied by the flow rate. The volume of oxygen consumed must be corrected to standard conditions, which are a barometric pressure of 760 mm Hg and a temperature of 0°C.

···

can modify heat loss and rate of metabolism by increasing or decreasing the insulation provided by the integument. This can be accomplished nearly instantaneously by the erection or compression of the feather or fur coat, thereby trapping or expelling air in the coat, and by increasing or decreasing peripheral blood flow, which functionally modifies the thickness of the integument and therefore its thermal permeability. Heat loss is also proportional to the effective surface area of an endotherm and therefore is affected by posture. Insulation can be modified on a seasonal basis in fall and spring as the feathers or fur are replaced, when the thickness of the coat can be changed.

As a result of compensatory changes in insulation, peripheral circulation, and posture, the rate of metabolism in an endotherm remains constant over a range of ambient temperatures (fig. 1.2B; see fig. 1.1), even though ΔT changes over that range. This range of temperatures is called the *zone of thermoneutrality* (see fig. 1.1). The rate of metabolism measured in an adult animal that is regulating its body temperature within the zone of thermoneutrality when it is inactive during the inactive period and postabsorptive (i.e., when it is not digesting a meal) is called the *basal rate of metabolism*

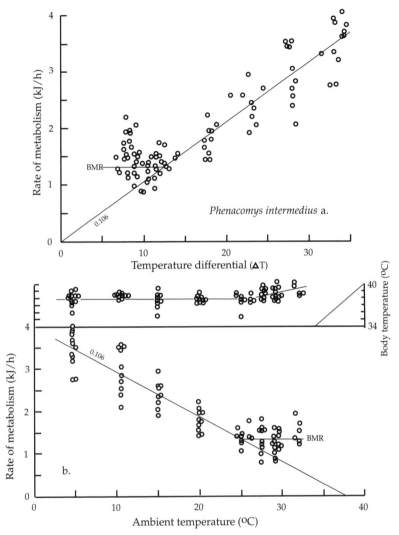

Figure 1.2. (A) Rate of metabolism as a function of the differential between body temperature and ambient temperature in mountain phenacomys, *Phenacomys intermedius*. (B) Rate of metabolism and body temperature as a function of ambient temperature in *Phenacomys intermedius*. (Modified from McNab 1992a.)

(or BMR) because that is the lowest rate of metabolism normally compatible with temperature regulation (McNab 1997).

The basal rate is often used to characterize endotherms, not because an individual spends much time in the conditions required by the definition of BMR, but because the conditions under which BMR is measured are, by definition, the same for all endotherms. Therefore, BMR is an equivalent measure of energy expenditure in all endotherms, unlike measurements at some fixed ambient temperature or on free-living animals in the field. Uniformity in the definition of BMR permits the relationships of this rate to the ecology, behavior, and distribution of endotherms to be examined—an effort that has had fruitful results (see chaps. 3–10). Furthermore, mass-independent variations in the field energy expenditures of endotherms are correlated with mass-independent variations in BMR (see chap. 11), which gives BMR greater significance than would be normally expected from laboratory measurements.

Speakman et al. (1993) argued that it is nearly impossible to ensure that measurements on many species are made on animals in their "basal" state. They also maintained that Kleiber's (1961) definition of BMR did not include the stipulation that body temperature was constant or that the measurements were to be made at some period in the daily cycle. Both statements are correct, but that does not mean that basal rate of metabolism, as a practical matter, cannot be effectively used, if we add the caveat, as most people do, that it applies only to endotherms when regulating their "normal" body temperature at the inactive part of their daily cycle. This may be an extension of Kleiber's criteria, but it presents no clear problem. Anyone who has measured energy expenditure is fully aware of the difficulties of attaining a resting state, but several conditions should be required, including being in the zone of thermoneutrality, which means that one measures the rates over a range of ambient temperatures during the period of rest and when the animal is maintaining its normal body temperature. (Sometimes investigators, for simplicity, arbitrarily choose a temperature as being in thermoneutrality [e.g., 30°C by Wiersma et al. 2007], but that decision has risks, especially when applied to species of all sizes.) Of course, none of these criteria evades the difficulty of measurements taken in artiodactyls that use gut fermentation; it may be impossible to get postabsorptive values without injuring the animal.

Stephenson and Racey (1995) argued that the use of BMR is inappropriate in species that enter torpor because it gives values that are unrealistically low, especially in some insectivores. They refer to all measurements of shrews and tenrecs as "resting" rates because of the tendency of some of these species to enter torpor. This view has been followed by Symonds (1999), but presents a problem and misinterprets an observation. The problem is that the use of a "resting" rate means that the rate is not equivalent in all species, which makes species comparisons subject to arbitrary decisions. The misinterpretation, as we shall see, is that nearly all birds and mammals that enter torpor have low BMRs, even when maintaining their normal body temperature (see chaps. 3, 4, and 8). The capacity to use daily torpor is another factor that determines BMR: species that enter torpor do not have the same BMR as species of the same mass that do not enter torpor.

At ambient temperatures below the zone of thermoneutrality, the rate of metabolism of endotherms increases (see fig. 1.2B) because ΔT has increased to the point where changes in insulation, posture, and peripheral circulation can no longer compensate for the increased heat loss dictated by ΔT. The rate of metabolism increases at temperatures below thermoneutrality as long as ΔT increases, or better stated, ΔT increases as long as the rate of metabolism adequately increases with a fall in T_a. When the curve of rate of metabolism plotted on T_a below thermoneutrality extrapolates to zero metabolism at $T_b = T_a$, its slope equals thermal conductance (see fig. 1.1), which is the inverse of insulation. Thus, the lower limit of thermoneutrality is an ambient temperature that separates the responses of endotherms at higher temperatures—by changes in insulation, posture, or peripheral circulation (the region of "physical" thermoregulation) from their responses at lower ambient temperatures, which require a change in rate of metabolism (the region of "chemical" thermoregulation).

At least that is how things are supposed to be in this idealized relationship, a condition most often seen in small species. Large species, however, often do not sharply distinguish the ambient temperatures at which physical and chemical thermoregulation occur (McNab 1980a). When that is the case, the metabolism-temperature curve below thermoneutrality usually extrapolates to $T_a > T_b$, and its slope is not a measure of thermal conductance and insulation (dashed curve, fig. 1.3). Under these circumstances, the curve below thermoneutrality is normally broken into a series of curves, each of which extrapolates to the mean T_b that corresponds to the appropriate curve, as is seen in data from the Auckland Island flightless teal (*Anas a. aucklandica*). Under this condition, conductance decreases with a decrease in ambient temperature below thermoneutrality until a minimal conductance is attained.

The ability of an endotherm to maintain a constant body temperature is limited at both low and high ambient temperatures. The increase in rate of metabolism with a decrease in T_a continues until a limit to ΔT is reached, which usually defines the minimal T_a that can be tolerated. This means that a further increase in rate of metabolism with the further decrease in ambient temperature is inadequate to maintain the additional increase in ΔT or that an increase in insulation does not compensate for the increased heat loss. Below this limit, body temperature decreases.

At high ambient temperatures, heat loss must increase to prevent overheating. Overheating at high ambient temperatures can be avoided by decreasing insulation, increasing peripheral circulation, and principally by increasing evaporative water loss. Unless adequate evaporation of water occurs

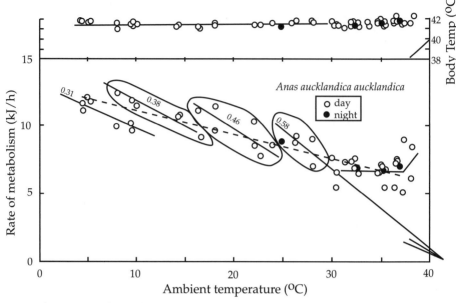

Figure 1.3. Rate of metabolism and body temperature as a function of ambient temperature in the Auckland Island flightless teal (*Anas aucklandica aucklandica*). The indicated slopes of the solid lines below thermoneutrality represent estimates of thermal conductance, whereas the slope of the dashed curve is not thermal conductance because it does not extrapolate to body temperature at zero metabolism. (Modified from McNab 2003a.)

at high ambient temperatures, heat will be stored, and the body temperature and rate of metabolism will increase, which threatens heat stroke if the increase in body temperature continues. Heat loss at cold ambient temperatures and heat storage at high ambient temperatures, then, define the limits of the zone of thermoneutrality.

Many of these relationships are summarized by the simplistic, but informative, Scholander-Irving equation:

$$M = C(T_b - T_a), \tag{1.1}$$

where M is rate of metabolism (kJ/h), C is thermal conductance (kJ/h · °C), T_b is body temperature (°C), and T_a is ambient temperature (°C) (Scholander et al. 1950b). The ability of an endotherm to maintain a temperature differential with the environment, $\Delta T = T_b - T_a$, is proportional to the ratio of the rate of heat production to thermal conductance, M/C, or alternatively, to the product of heat production and insulation, $M \cdot I$, where $I = 1/C$. This relationship is simplistic in the sense that evaporative heat loss is ignored, as are heat exchange with a radiational source, such as the sun or a cold sky, and convective exchange in terrestrial or aquatic environments.

These complications have been dealt with elsewhere (Porter & Gates 1969; Tracy 1972; Gates 1980). These "physical" topics are obviously important, but usually under restricted environmental conditions.

All four components of the relationship described by equation (1.1), M, C, T_b, and T_a, vary. T_a depends on the environment in which a species lives, time of day, and season. The other three components characterize and vary with species, and it is the exploration of these components that has given rise to our (limited) understanding of the energetics of endotherms, which is the basis of this book and its attempt to explore the consequences that these variations have for endotherms. We will examine the variation in each of these terms. We begin by examining what determines the basal rate of metabolism, and we will see how it can be used as a standard for comparing the performance of various endotherms. The same approach will be used to examine the factors that influence thermal conductance and body temperature.

The impact of body mass on basal rate of metabolism

Various factors influence the BMR of endotherms, but by far the most important is body mass because it has a numerical range much greater than any other factor. The most prevalent means of describing the relationship between BMR and body mass is as a power function:

$$\text{BMR (kJ/h)} = a \cdot m^b, \tag{1.2}$$

where BMR, expressed in terms of kilojoules per hour, equals the product of a coefficient, a, which has units of $kJ/g^b \cdot h$, and body mass, m, in grams, raised to a dimensionless power b, which is almost always <1.00. This relationship is most easily estimated by fitting the data to a logarithmic transformation:

$$\log_{10} \text{BMR} = b \cdot \log_{10} m + \log_{10} a. \tag{1.3}$$

In this form, the power b is the slope of the curve of $\log_{10}$ BMR plotted on $\log_{10} m$. This analysis can be done on all available data, on a selected subset of available data, such as members of a particular genus, family, or order, or on data organized by other factors, depending on the question of interest.

Sarrus and Rameaux (1839) examined heat production in endotherms. They argued that the surface area ultimately determined heat loss and therefore heat production, a conclusion that may have led to Bergmann's (1847) argument that mammals get larger in cold climates because surface-to-volume ratio and heat loss decrease with an increase in mass (see chap. 6).

This view was codified by Rubner (1883) with his "surface law," where rate of metabolism was described as proportional to $m^{0.67}$. However, several problems exist with this "rule." A fundamental one is that von Hoesslin (1888) showed that rate of metabolism in fish scales close to surface area, a relationship that is inappropriate in ectotherms. Another problem has been the question as to what surface area, given behavioral and circulatory changes in the functional surface area.

These difficulties were put aside by Kleiber, who in 1932 argued that the rate of metabolism in mammals is actually proportional to $m^{0.75}$. As was clearly stated by Schmidt-Nielsen (1970): "We were . . . relieved of the constraining demand to fit metabolic rate to body surface, and the heated discussions of how to determine 'free' or 'true' surface have therefore subsided. The 'surface law' as such does not even survive as a 'surface rule,' but the analysis of function in relation to body size has in itself become an interesting and productive field." (Yet, explanations that fall out of favor have a tendency to return in another form; see Glazier [2005, 2008].)

An extended discussion of causes for the observation that $b < 1.00$ has raised two questions: (1) Why is $b < 1.00$? and (2) What is the "real" value for b? The value and "meaning" of b in mammals has been extensively discussed. Some investigators have argued that a "universal" b exists and that it is (Heusner 1991), or may be (White & Seymour 2003), 0.67, whereas others agree that a universal b exists, but maintain that it is 0.75 (MacMahon 1973; West et al. 1997, 1999; Banaver et al. 1999, 2002; Gillooly et al. 2001; Savage et al. 2004). A difficulty with these arguments is that they invariably ignore the residual variation in BMR. That leaves some doubt whether a fixed, universal b exists (Bokma 2004; Glazier 2005; White et al. 2007b, 2009), especially if the residual variation is associated with factors that correlate with mass, a situation that will cause any fitted b to vary as these factors are added to, or dropped from, an analysis (McNab 2008a, 2009a).

This analysis raises additional questions. One question concerns the possibility that b is not constant, but curvilinear. Kolokotrones et al. (2010) showed that at small masses and at very large masses, the basal rate of metabolism is higher than expected from a curve in which b is linear. This observation leads to a curvilinear b. The principal difficulty with this analysis is that a fitted b assumes that the distribution of species along such a curve has a set of equal characteristics so that no bias is given by factors other than body mass. However, the data usually used for an estimate of b provide no assurance that this criterion was met. For example, at the smallest masses, shrews that belong to the genus *Sorex* and arvicoline rodents have unusually high mass-independent BMRs (see chap. 4). At large masses, terrestrial and aquatic carnivores tend to have high mass-independent BMRs, which is why the fitted curvilinear curve extrapolates to the orca (*Orcinus orca*), even though it was not used in the analysis. In contrast, quite a few species of intermediate mass, such as armadillos, sloths, and marsupials, have low BMRs. The frequency distribution of mass-independent BMRs therefore produces a distribution that is high at small masses, low at intermediate masses, and high at large masses. Kolokotrones and colleagues tried to compensate for this distribution by adding a temperature coefficient to the analysis, but that confuses endotherms with ectotherms, as will be noted later. So, doubt exists concerning the existence of a curvilinear b.

Another question is whether the power b is constant or variable, especially given the disagreement on the values of b. Glazier (2005, 2008, 2010) proposed a hypothesis that b varies with the level of energy expenditure: surface area constraints on heat and resource fluxes are reflected in $b = 0.67$, and when surface area constraints are not present, energy expenditure is limited and b approaches 1.00. If this is the case, then b should vary as a function of energy expenditure, which depends on the balance between heat and resource flux and energy use. This variation is demonstrated in mammals in figure 1.4A, where b for various metabolism-mass curves is plotted as a function of the mean rate of metabolism (and therefore the "height" of the curve) at a fixed mass of 50 g. Notice that the pattern is U-shaped: hibernating and maximally active individuals have b equal to, or approaching, 1.00, as expected in individuals in which the limit to energy expenditure depends principally on body mass, whereas resting, field active, and torpid individuals have b similar to 0.67, which implies that heat loss is proportional to surface area. A similar U-shaped pattern occurs in birds (fig. 1.4B). These patterns support the view that the power of body mass that determines BMR reflects multiple constraints and may contribute to the appearance of curvilinearity in b.

And then there are the residual variations in BMR. In mammals, BMR at any particular mass varies by a factor between 3:1 and 10:1; in birds, the variation is less, but it still ranges from 2:1 to 5:1. This variation suggests that factors other than body mass influence BMR in these endotherms, as will be shown in chapter 3, and that the variation is not simply "measurement error." Indeed, as we shall see, the restricted variation in birds, as compared with mammals, reflects the usual absence in birds of habits that depress BMR, which in mammals include commitments to fossorial, arboreal, or inactive lifestyles. In birds, a reduced BMR is principally associated with the loss of flight and a sedentary lifestyle. Furthermore, body composition may influence b: "metabolically active organs constitute a smaller percentage of body mass in larger size mammals" (Wang et al. 2001). The impact of residual variation in basal rate of metabolism can be examined using the dimensionless, mass-independent BMR:

$$\text{mass-independent BMR (\%)} = 100(\text{measured BMR [kJ/h]}/ \text{calculated BMR [kJ/h]}),$$

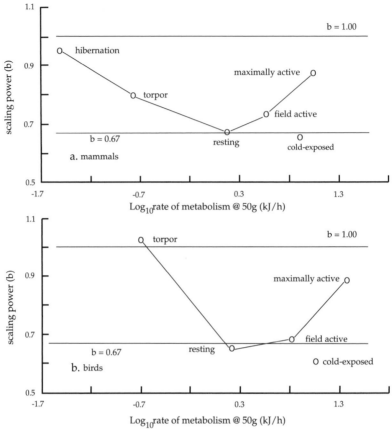

Figure 1.4. Metabolic scaling power of body mass as a function of the level of metabolism in (A) mammals and (B) birds at a mass of 50 g in various physiological states. (Modified from Glazier 2008.)

where calculated BMR is derived from the appropriate version of equation (1.2).

Because basal rates of metabolism are proportional to m^b, mass-specific rates are proportional to $m^b/m^{1.00} = m^{b-1.00}$. Thus, if $b = 0.75$, then the mass-specific rate is proportional to $m^{-0.25}$. Indeed, most reported measurements are in terms of mL O_2/g · h. However, the fundamental relationship is between total rate of metabolism—after all, that is what is measured—and body mass (McNab 1999) (see box 1.2). The principal justification for the use of mass-specific rates, besides custom, is that they represent turnover times, which may be of value under some circumstances, but the rate that most animals are concerned with is the amount of food required. Elephants do not know that they have low mass-specific rates of metabolism: they must spend most of the day eating, reflecting m^b.

BOX 1.2. ···

The Search for Simple Biological Generalities and a Defense of Natural History

Biologists can be classified by their many approaches—ecological, behavioral, evolutionary, physiological, cellular, molecular, and so forth—all of which must be interconnected. An interesting subset of biologists searches for general principles from a mathematical viewpoint. This approach is valuable in itself and often guides empirical biologists in making meaningful observations. However, theoretical biologists often appear to be seduced by the success of theoretical physicists in finding the, or an, equation that will describe, or "explain," a broad range of phenomena. Thus, if we could find *the* proper mass equation, it would account for all of the variation in some quantitative relationship. However, an appreciable amount of residual variation exists around all biological relationships. This variation usually reflects the ability of organisms to take advantage of the loopholes that exist in (nearly?) all biological relationships. The relationships are therefore modified by a set of conditional clauses: "if . . . , then . . . ." In fact, the exploitation of conditional clauses is what has produced much of biological diversity. This does not mean that mathematical descriptions of biological relationships are not valuable; after all, we shall see how valuable they are as a standard against which to evaluate species performance. However, they are usually more complex than they were originally thought to be; a good example is the Kleiber metabolism-mass curve (see chap. 3). Only when the residual variation in biological phenomena is included can we say that we understand the true impact of those phenomena. And this is where natural histories are essential: without knowledge of where species live, how they behave, and what they eat, we have no understanding of the basis for residual variation. Without natural histories, a species is simply a dot on a graph.

···

Mammals

An analysis of data on 639 species of mammals (McNab 2008a) indicated that BMR in mammals correlates with body mass in the following relationship:

$$\text{BMR (kJ/h)} = 0.070 \, g^{0.721}, \tag{1.4}$$

when the correlation coefficient $r = 0.984$ and $r^2 = 0.968$, which indicates that 96.8% of the variation in BMR is accounted for simply by variation in body mass (fig. 1.5A). This high r^2 derives from the great range in body mass among the measured species—namely, from a 2.2 g shrew (*Suncus etruscus*) to the 3,221,000 g orca, a $10^{6.17}$-fold range in mass. No factor other than mass that potentially influences the BMR of mammals has such a great range of states, which accounts for its great impact. Equation (1.4) is used as a mass standard for BMR in mammals in this book. When used as a standard for measurements in terms of oxygen consumption (mL O_2/h), the

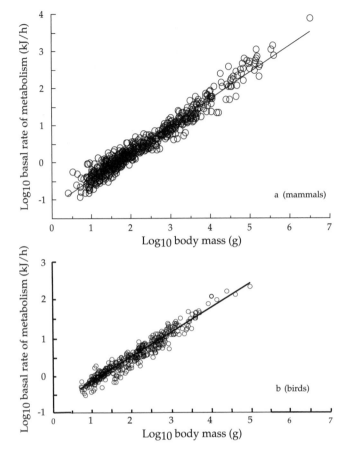

Figure 1.5. Log_{10} basal rate of metabolism as a function of log_{10} body mass in (A) mammals and (B) birds. (A modified from McNab 2008a; B modified from McNab 2009a.)

coefficient 0.070 is multiplied by 50 (mL O_2/kJ) $= 1/0.02$ kJ/mL O_2; then $a = 3.50$ (mL O_2/$g^{0.721}$ h).

Birds

A similar pattern is found in birds. In 533 species of birds, the BMR is described by

$$\text{BMR (kJ/h)} = 0.145 \ g^{0.652}, \tag{1.5}$$

when $r^2 = 0.942$ (McNab 2009a). Thus, 94.2% of the variation in avian BMRs is accounted for by variation in body mass (fig. 1.5B), which also reflects the large range of body mass among measured species—namely, from a 5.2 g sunbird (*Aethopyga christinae*) to the 90,000 g ostrich (*Struthio camelus*), a $10^{4.24}$-fold range in mass. (The smaller r^2 in birds reflects the

85.1-fold smaller mass range in measured birds than in measured mammals.) The coefficient in this equation, when converted to oxygen consumption, equals $0.145 \times 50 = 7.25$ (mL $O_2/g^{0.652}$ h).

Comparing equations (1.4) and (1.5) indicates that birds generally have higher BMRs than mammals of the same mass (i.e., by a factor of $[0.145/0.070] = 2.07:1$ at a mass of 1 g). However, because the power of mass in birds is smaller than that in mammals, this difference gradually diminishes with an increase in mass. For example, at 1 kg, the ratio of bird to mammal BMR is $(13.10/10.19) = 1.48:1$, and it is $(58.80/53.59) = 1.10:1$ at 10 kg. A mass therefore exists at which the mean BMR of birds equals that of mammals: it is about 38.3 kg (fig. 1.6). Only a few living birds exceed that mass: the ostrich (*Struthio camelus*), emu (*Dromaius novaehollandiae*), and southern cassowary (*Casuarius casuarius*), all of which are flightless. Residual variation in BMR exists in both relationships, so that some species

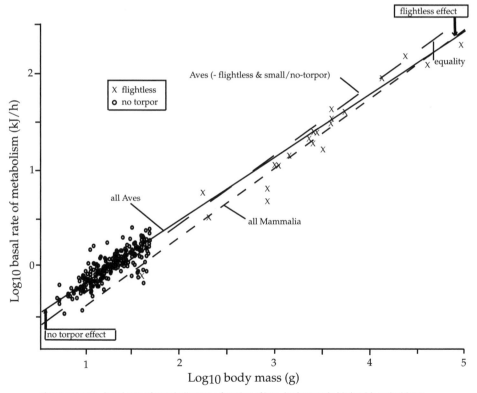

Figure 1.6. Log₁₀ basal rate of metabolism as a function of log₁₀ body mass in birds with and without species with masses less than 50 g that enter torpor and species that are flightless. Also indicated are fitted curves for all birds (McNab 2009a) and for all mammals (McNab 2008a). (Modified from McNab 2009a.)

of birds at almost all masses have BMRs equal to, or lower than, some mammals of the same mass, reflecting the fact that body mass does not account for all the variation in BMR in these two groups. Factors other than mass that appear to influence the BMRs of birds and mammals are extensively explored in chapter 3.

The equality of basal rates in birds and mammals reflects the lower scaling power b in birds, which results from two factors. One is the prevalence of a flightless condition in 9 of the 14 species that weigh more than 3.0 kg, which drags down the bird curve at large masses. A second is that most small birds are committed to continuous endothermy, which raises the bird curve at masses <50 g (McNab 2009a): birds that enter torpor have basal rates that are 78% of those of birds that do not use torpor. Many more similarly sized mammals enter torpor and collectively have lower basal rates (81% of non-torpor species; see chap. 4) than birds. For example, temperate bats, most of which respond to winter by hibernating in local caves, feed on flying insects, and the few migratory species have low basal rates (Genoud 1993). In contrast, temperate insectivorous birds that feed in flight migrate to warm-temperate or tropical environments to avoid a shortage of food in winter, thereby avoiding the obligatory use of torpor. When flightless birds and birds with masses <50 g that enter torpor are deleted from the bird curve, it is parallel to the mammal curve:

$$\text{BMR (kJ/h)} = 0.105 \; m^{0.708}. \tag{1.6}$$

Now the bird curve has a coefficient that is 50% greater than the mammal curve ($0.105/0.070 = 1.5$ [eq. (1.4), fig. 1.6]). The principal reason why birds have higher basal rates than mammals appears to be their expensive form of flight (McNab 2009a): sedentary birds have lower BMRs than migratory species, and flightless species have eutherian-like BMRs (see chap. 8).

The impact of body mass on minimal thermal conductance

As with basal rate of metabolism, body mass is the most important factor influencing minimal thermal conductance. Thermal conductance in birds and mammals is highly variable. As we have seen, the lower limit of thermoneutrality in small species is usually defined by minimal thermal conductance—that is, maximal insulation. Intermediate and large species, however, have conductances that often continue to decrease with ambient temperature and reach a minimum only at ambient temperatures below the lower limit of thermoneutrality. So, rather than trying to predict the various circumstances that influence the minimal conductance, which would turn out to be a complicated goal, we are usually concerned with the conductance that defines the lower limit of thermoneutrality.

Mammals

Thermal conductance in mammals is principally defined by body mass. Morrison and Ryser (1951), Morrison and Tietz (1957), and McNab and Morrison (1963) made the first description of this relationship. Since then, Herreid and Kessel (1967) and Bradley and Deavers (1980) have reviewed the literature. All these descriptions were similar. The most recent review of the literature by Aschoff (1981) indicated that for 59 species during rest, when conductances were lower than during activity (fig. 1.7),

$$C \,(\mathrm{kJ/h} \cdot {}^{\circ}\mathrm{C}) = 0.020 \; m^{0.481}. \tag{1.7}$$

Birds

Lasiewski et al. (1967) described thermal conductance in 35 species of birds:

$$C \,(\mathrm{kJ/h} \cdot {}^{\circ}\mathrm{C}) = 0.017 \; m^{0.492}. \tag{1.8}$$

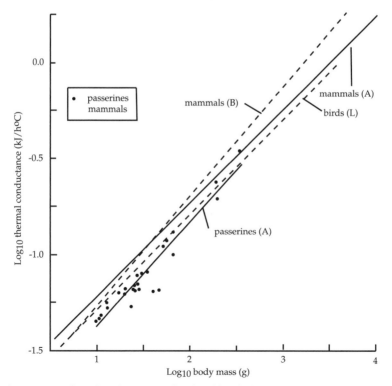

Figure 1.7. Log_{10} thermal conductance as a function of log_{10} body mass in mammals and passerine birds. (Data from Aschoff 1981 [A], Bradley & Deavers [B], and Lasiewski et al. 1967 [L].)

Aschoff (1981) separated passerines from other birds (see fig. 1.7). He indicated that for 37 passerines during rest, when conductances were lower than during activity,

$$C\,(\text{kJ/h} \cdot {}^\circ\text{C}) = 0.012\ m^{0.539},\qquad (1.9)$$

and for 11 nonpasserines,

$$C\,(\text{kJ/h} \cdot {}^\circ\text{C}) = 0.019\ m^{0.417}.\qquad (1.10)$$

Avian conductances are lower than mammalian conductances. Herreid and Kessel (1967) showed that lizards had thermal conductances that are about 10 times those of mammals and birds.

Another view of the difference in thermal conductance between birds and mammals is that it reflects a difference in body temperature. Birds have higher body temperatures and lower conductances than mammals except at large masses, at which the two groups are equal. A reduction in body temperature is associated with a decrease in blood circulation, which undoubtedly leads to a decrease in peripheral circulation and therefore in heat loss. In this case, a reduction in thermal conductance is not independent of a decrease in rate of metabolism, but is rather its physiological by-product. Clearly, change in thermal conductance is a component of a dynamic thermal system.

Climate

Climate also affects thermal conductance, but its impacts have not been numerically evaluated. Scholander et al. (1950a,b,c) argued that changes in insulation were the principal response to climate and to seasonal changes in the thermal environment. Mammals that live in cold climates, such as those at high latitudes and altitudes, have lower thermal conductances than species of similar masses that live at low latitudes and altitudes. Thus, naked or nearly naked mammals, including pangolins and armadillos, are usually limited to the tropics, although the nakedness of some of the largest mammals, rhinoceroses and elephants, appears to be a means of reducing overheating in a hot climate. Indeed, cold-climate rhinos and mastodons were heavily furred (see chap. 6).

Season

Season influences thermal conductance, but comparatively little effort has been directed toward this subject. Seasonal variation in conductance (and insulation) is widespread in temperate and polar environments; it is what makes the winter pelts of furbearers more valuable than their summer pelts. Hart (1956) demonstrated a correlation of insulation with the thickness of the fur coat in mammals. Of course, an increase in the thickness of the fur or feather coat can occur only in species large enough to carry a thick insulative layer. Thus, an appreciable decrease in conductance in winter occurs in snowshoe hares (*Lepus americanus*) (Hart et al. 1965), and presumably in larger species, but no such decrease is apparent in cold-acclimatized white-footed mice (*Peromyscus leucopus*) (Hart & Heroux 1953), although a small increase in fur length and density has been seen in *P. maniculatus* in relation to an altitudinal distribution (Wasserman & Nash 1979). The demonstration by Kendeigh et al. (1977) that most birds have a lower limit of thermoneutrality in winter than in summer implies seasonal variation in thermal conductance.

Compensations for variation in BMR

Minimal conductance (maximal insulation) cannot be taken out of its thermal context. An appreciable decrease in conductance in some cases appears to compensate for a reduction in the rate of metabolism. Desert species and populations of *Peromyscus* have lower conductances than mesic species and populations (McNab & Morrison 1963), so that the ratio BMR/C (see eq. [1.1]) would remain unchanged in a perfect system, which would permit a desired ΔT and T_b to be maintained. Thus, *Peromyscus* from mesic areas have BMR/C ratios from 0.93 to 1.29, those from desert mountains from 0.93 to 1.11, and those from desert lowlands from 1.00 to 1.05 — that is, nearly equal adjustments to life in a diversity of climates, even though BMRs were reduced in desert populations by as much as 27% (see chap. 7). However, this numerical adjustment works only when the minimal conductance defines the lower limit of thermoneutrality, which generally occurs in small species.

A low thermal conductance is also found in some mammals that live in tropical rainforests, which might be unexpected. For example, the binturong (*Arctictis binturong*) has a BMR that is only 33% of the value expected from its 15 kg mass (McNab 1995). In compensation, the binturong has a very thick fur coat, whose minimal conductance is only 13% of the value expected from the animal's mass — something that obviously would not be expected in a lowland tropical species! Consequently, the ΔT that could be maintained at the lower limit of thermoneutrality would be $0.33/0.13 = 2.5$ times that of a standard mammal. However, the relationship between rate of metabolism and ambient temperature in the binturong does not conform to the usual pattern found in small species (fig. 1.8), so the minimal conductance does not define the lower limit of thermoneutrality, but occurs at very low ambient temperatures. The BMR/C ratio at the lower limit of thermoneutrality, at 27°C, is actually $0.33/1.03 = 0.32$, or only one-third the value expected

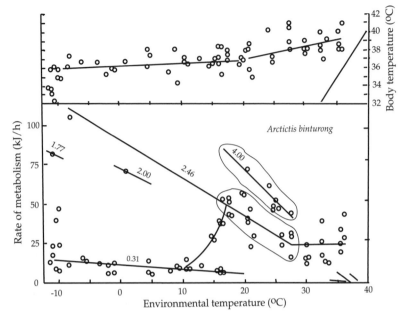

Figure 1.8. Rate of metabolism and body temperature as a function of ambient temperature in the binturong (*Arctictis binturong*). (Modified from McNab 1995.)

in a standard mammal, but as T_a falls, C decreases to its minimal value. A similar response is found in tree sloths (McNab 1978c) and in the tropical, frugivorous carnivorans *Arctogalidia trivirgata* and *Nandinia binotata* (McNab 1995). This behavior occurs principally in mammals with low BMRs at masses from 1 to 14 kg, but it is also found in some pteropodid bats that weigh more than 500 g (McNab 2002c). These observations emphasize that rate of metabolism and thermal conductance are part of a coordinated system.

Body temperature

The thermal context of the body temperature of endotherms must be considered as well. Variations in the body temperature of ectotherms obviously influence the rate of metabolism according to a Q_{10} relationship: rates of metabolism increase or decrease by a factor of about 2 to 3 for each 10°C increase or decrease in body temperature. This analysis has been applied to endotherms as well (Gillooly et al. 2001; White & Seymour 2004; Clarke et al. 2010), as has the analysis of Kolokotrones et al. (2010), but the analogy of birds and mammals with ectotherms is questionable.

Given the basis of temperature regulation in endotherms, a reverse causality is appropriate: endotherms have high or low body temperatures *because* they have high or low rates of metabolism, not vice versa. The heat content and body temperatures of endotherms are determined by the ratio M/C: $T_b = (M/C) + T_a$ (see eq. [1.1]). Thus, a low conductance can be coupled with an average rate of metabolism to produce a slightly high body temperature, and a high conductance means that a high rate of metabolism is required to maintain a normal body temperature. Birds generally have higher body temperatures than mammals *because* most birds have both higher rates of metabolism and lower conductances than most mammals. The dependence of body temperature on the rate of metabolism is the reason why there are no "cool" terrestrial endotherms. Such a strategy might be justified on theoretical grounds: energetic economy would be gained by an Arctic mammal or bird that regulated its body temperature at 20°C–25°C, rather than 37°C–42°C, because ΔT and heat loss would be appreciably reduced, but the capacity for temperature regulation and the maintenance of a large ΔT requires a high rate of metabolism and a low conductance.

The opposite extreme can be seen in the naked mole-rat (*Heterocephalus glaber*) from East Africa (fig. 1.9): it has a BMR that is 53% of the value expected from mass coupled with a conductance that is 242% of the value expected from mass, so that the mass-independent ratio BMR$/C = 0.53/2.42 = 0.22$ — only 22% of the value expected for a standard 40 g mammal (McNab 1966, 1979b). Consequently, this species cannot tolerate low ambient temperatures and is capable of maintaining a maximal ΔT that is only 7.5°C; body temperature falls to 12°C when $T_a = 10$°C. This combination of a low rate of metabolism and a very high thermal conductance is a response to life in a warm, humid burrow system (see chap. 7), which has nearly converted *Heterocephalus* into a poikilotherm. In spite of its thermoregulatory incompetence, *Heterocephalus* is not an ectotherm: no ectotherm increases its rate of metabolism as T_a decreases, as *Heterocephalus* does. Whether *Heterocephalus* is a poikilotherm, as has been argued by Buffenstein and Yahav (1991), is unclear because poikilothermy and incompetent homeothermy may form a graded series; it is, however, an endotherm.

The condition of *Heterocephalus* should not be confused with entrance into torpor or hibernation, which are voluntary conditions from which endotherms can spontaneously arouse (see chap. 8). *Heterocephalus* cannot spontaneously arouse from low body temperatures, which is not a practical problem because this species never encounters low ambient temperatures in its tropical subterranean environment. Burrow temperatures are usually about 30°C. As evidence of the unnatural occurrence of low body temperatures in this species, its body fat reserves congeal when T_b falls below about 20°C and "melt" only when an appropriate body temperature is restored (artificially, as when the animal is held in one's hand), permitting the individual to move again.

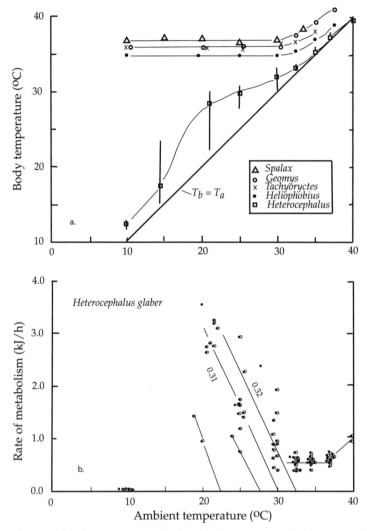

Figure 1.9. (A) Body temperature as a function of ambient temperature in various fossorial rodents. (B) Rate of metabolism as a function of ambient temperature in the naked mole-rat, *Heterocephalus glaber*. (Modified from McNab 1966.)

Other endotherms

Birds and mammals are not the only endotherms on earth, but the endothermy of other groups is limited in one way or another. Among vertebrates, an intermediate form of endothermy is found in some fishes, including large, active sharks and tunas. Their endothermy is usually limited to their lateral muscles. In most "cold" fishes, the heat produced by muscle contraction is transported by blood to the heart and then to the gills, where it is lost through gas exchange, so little temperature differential with the water is maintained in the muscles (no more than 1°C–2°C). In contrast, these "hot" fishes shunt venous blood from muscles through a countercurrent heat exchanger that transfers the majority of the heat produced in the muscles to warm arterial blood flowing to the muscles. This mechanism can maintain a ΔT of up to 20°C at a water temperature of 7°C in a 200–350 kg bluefin tuna (*Thunnus thynnus*) (Dizon & Brill 1979). Smaller species, such as the yellowfin tuna (*T. albacares*), which usually weighs between 80 and 100 kg, maintain much smaller ΔTs, usually no greater than 4°C–8°C, and are restricted to warm waters (Barrett & Hester 1964; Dizon & Brill 1979). Some mackerel sharks (Lamnidae) also maintain temperature differentials in the lateral muscles of 6°C–8°C (Carey & Teal 1969; Carey et al. 1981; Smith & Rhodes 1983). The heat exchangers can be turned off in warm water and on in cold water, which permits effective control of body temperature.

Some sharks control the temperature of the stomach and liver, and they, along with rays (Mobulidae), swordfishes (Xiphiidae), and billfishes (Istiophoridae), control cephalic temperatures through the transfer of heat from the lateral musculature (Carey et al. 1971; Carey et al. 1982) or by means of cephalic heaters developed from eye muscles (Carey 1982; Block & Carey 1985; Block 1986; Alexander 1995, 1996). Most of these fishes are active predators whose endothermy enhances their predatory behavior, which pays its cost. For a general review of the evolution of endothermy in fishes, see Block et al. (1993) and Block and Finnerty (1994).

Female pythons (*Python molurus*) incubate their eggs by curling around them and maintaining a temperature differential of up to 7°C to 8°C with the environment (Van Mierop & Barnard 1978). They accomplish this by "shivering" (i.e., coordinated muscle contractions), which leads to an increased rate of heat production. That heat, combined with the reduction of exposed surface area and thermal conductance produced by the coiled posture, permits their eggs to develop more rapidly than they would otherwise. But as soon as incubation is finished, the female python returns to her normally ectothermic state. Male pythons do not show this behavior. In their normal ectothermic state, pythons often use basking behavior.

The leatherback turtle (*Dermochelys coriacea*), the largest marine turtle and one of the largest living reptiles (large adults usually weigh >500 kg), is often found in cold water. The question of whether these animals are endothermic revolves around the question of how to tell the difference between endothermy and inertial homeothermy: is a temperature differential with the environment a regulated ΔT, or is it a transitory result of an extended period of activity in an animal with a small surface-to-volume ratio? One measurement in Nova Scotia indicated a ΔT equal to 18°C in an active turtle at a water temperature of about 7.5°C (Frair et al. 1972), a ΔT that is unlikely to represent thermal inertia. No other marine turtle is found in cold water. Komodo monitors (oras, *Varanus komodoensis*) have body temperatures that are quite independent of ambient temperature, which reflects a mass that may reach 250 kg and a thick integument—that is, a temperature independence of 5°C–10°C based on thermal inertia (McNab & Auffenberg 1976; Auffenberg 1981).

The only other animals that are selectively endothermic are some insects. This condition most commonly occurs in large nocturnal moths, honeybees, bumblebees, euglossine bees, beetles, vespid wasps, and dragonflies (May 1979; Heinrich 1981). However, these insects are endothermic only when they are active. They arouse from an ectothermic state when inactive by "shivering" (i.e., by the rapid oscillation of their wings). The heat produced by this activity increases the temperature of the thorax, which is usually the only part of the body that is thermoregulated, although some heat is transferred to the abdomen to aid in the control of thoracic temperature. The restriction of endothermy in insects to the active state emphasizes the problem of paying the cost of continuous endothermy at a small body mass, as we shall see in chapter 4, although the collective activity of a bee or wasp colony can maintain a high colony temperature during the active season.

CHAPTER TWO. *Controversies in the Analysis of Quantitative Data*

Body mass obviously does not account for all the variation in the basal rates and thermal conductances of birds and mammals. Do factors other than mass influence thermal biology, or is residual variation simply measurement error? This question can be resolved by asking whether a pattern exists in residual variation: are the species that fall above the metabolism-mass curve more similar to one another in some manner than they are to those that fall below the curve? Indeed, many such patterns exist, and their exploration raises many questions. Why do some birds and mammals have higher BMRs than other species with the same body mass? Why do birds generally have higher BMRs than mammals? Do the habits and environments of species influence BMR? Does phylogeny influence character states independently of conditions encountered in the environment? If phylogeny is an important determinant of character states, how far back in generations does its determination go? These questions are not easily answered because the factors potentially determining phenotypic character states are intertwined. Efforts to untangle these interconnections have led to controversy over the relative importance of history, behavior, and the environment in the determination of character states.

The balance between the influences of environment and behavior and the influence of phylogeny on character states has been examined repeatedly. Most closely related species have similar body masses, behaviors, and food habits and live in similar environments, so their physiological similarity should not be surprising. Are they similar because they share a common ancestor, because they are restricted in their response to the environment, or because they live in similar environments and have similar behaviors? On the other hand, unrelated species that have similar masses, behaviors, and food habits and live in similar environments are often physiologically similar, and conversely, ecological outliers within a clade are usually physiologically distinct. Both observations suggest that body mass and phylogeny are not the only factors determining energy expenditure. Furthermore, phylogeny cannot anticipate accidents of history, such as the dispersal of birds and bats to oceanic islands or the selective evolution of a flightless condition in birds, occurrences that have had profound consequences for energy expenditure (see chap. 9).

One of the earliest attempts to explore whether the energetics of vertebrates correlates with conditions in the environment is found in the trio of articles written by Per Scholander, Laurence Irving, Raymond Hock, Fred Johnson, and Vladimir Walters in 1950, articles that might be considered to be the beginning of the field of physiological ecology, or at least of ecological energetics. These authors compared the rates of metabolism of birds and mammals from Barro Colorado Island, Panama, with those of species from Point Barrow, Alaska. Scholander and associates (1950b) concluded that *"the basal metabolic rate of terrestrial mammals from tropics to arctic is fundamentally determined by a size relation according to the formula Cal./day = 70 kg$^{\frac{3}{4}}$, and is phylogenetically nonadaptive to external temperature conditions. Equally nonadaptive is the body temperature, and the phylogenetic adaptation to cold therefore rests entirely upon the plasticity of the factors which determine heat loss, mainly the fur insulation"* [italics in original]. A serious problem with the design of this study was that the species studied in Panama were unrelated to the species studied in Alaska. This means that the similarities and differences found between these groups may not have been associated with the difference in environmental conditions. How does one get around this difficulty?

A comparatively simple solution, which was inspired by examining the Scholander study, is a comparison of close relatives that live in different environments or have different behaviors. McNab and Morrison (1963) compared the rates of metabolism in five species of *Peromyscus*, each of which had two populations found in different climates (see chap. 7). The populations in mesic coastal California had higher BMRs than the populations that lived in hotter, drier mountains, whereas those that lived in lowland deserts had the lowest BMRs (all rates were corrected for body mass, i.e., were mass-independent). Here, comparisons were made within and between species from the same genus and family. This study found a clear correlation between mass-independent BMRs and climatic conditions that was generally uncomplicated by evolutionary history, with the possible exception of one species, *P. californicus* (see chap. 7), a pattern seen later in the same species by Mueller and Diamond (2001).

Some species, however, have no close relatives. Thus, the red panda (*Ailurus fulgens*) has a low BMR (49% of the value expected from mass), but it is the only member of the genus *Ailurus* and is often classified as the only member of a unique family, the Ailuridae. This semiarboreal species feeds principally on bamboo shoots, grass, fruit, and acorns. However, the red panda can be compared with other members of the order Carnivora that have similar arboreal and trophic habits, including *Arctogalidia, Paradoxurus, Nandinia,* and *Arctictis,* all of which turn out, like *Ailurus,* to have low BMRs (McNab 1988a, 1995, 2005c).

This arboreal pattern has also been found among close relatives. Terrestrial macropods have BMRs that range between 82% and 96% of the value expected from the all-mammal curve, whereas an arboreal tree kangaroo, *Dendrolagus matschiei,* has a BMR equal to 71% of that value (McNab 1988a). The arboreal margay (*Leopardus wiedii*) has a BMR equal to 73%; its close relative, the terrestrial ocelot (*L. pardalis*), has a BMR equal to 114% (McNab 2000c). These arboreal mammals, including the red panda, have low mass-independent BMRs regardless of their evolutionary affiliation. A convergent pattern in metabolism is also found among the diverse mammals that are committed to feeding on ants and termites (see chap. 5).

Another way to examine the factors that determine the energetics of endotherms is to look for ecologically, behaviorally, or environmentally distinct species and compare them with their "conventional" relatives. As we will see in chapter 9, fruit pigeons that belong to the genus *Ducula* and live only on small oceanic islands have mass-independent BMRs that are 68% of the values for *Ducula* committed to life on the large island of New Guinea (McNab 2000b); a similar pattern is found in flying foxes (Pteropodidae) (McNab & Bonaccorso 2001). An insectivorous leaf-nosed bat (*Macrotus californicus*) of the subfamily Phyllostominae, family Phyllostomidae, has a BMR equal to 70%. Other members of this subfamily include three omnivorous species of the genus *Phyllostomus,* which have BMRs between 78% and 85%; a species (*Tonatia bidens*) that consumes vertebrates and insects has a BMR of 104%, and a dedicated carnivore (*Chrotopterus auritus*) has a BMR of 109% (McNab 1969, 2003b). These examples illustrate how species showing significant deviations in behavior can have energy expenditures that are higher or lower than those of their "conventional" relatives.

An approach that attempts to formally "correct" for the impact of history is the use of "independent contrasts," a technique that incorporates a cladogram of the species in the analysis. The principal justification for its use is a concern with analyzing the occurrence of character states without taking history into account because their acquisition often is not independent of phylogeny (Felsenstein 1985). As justified as this concern is, several difficulties exist with the use of independent contrasts, or any other phylogenetic technique that does not take environmental conditions into account. An important one is that these techniques essentially give priority to the influence of "history" (McNab 2003b, 2009b) without addressing the impact of a clade's characteristics. With reference to BMR, the phylogenetic approach comes close to stating that a particular clade can have any BMR dictated by historical events because its level is not associated with any factor other than body mass and phylogeny. This view essentially denies the action of natural selection on *contemporary* organisms.

Phylogenetic analyses presumably depend on the accuracy of the clado-gram used. However, all independent contrast analyses of the evolution of the energetics of birds have been based on the Sibley-Ahlquist cladogram (1990), including those by Reynolds and Lee (1996), Rezende et al. (2002), McKechnie and Wolf (2004), and Wiersma et al. (2007). This technique led them to conclude that passerines do not have basal rates greater than the collective of other birds, whereas Lasiewski and Dawson (1967), As-choff and Pohl (1970a,b), Kendeigh et al. (1977), and McNab (2009a), using analyses of covariance, argued that passerines do in fact have higher basal rates. Recent work on DNA base sequences in birds has indicated that the Sibley-Ahlquist cladogram is highly misleading (Barker et al. 2004; Fain & Houde 2004; Ericson et al. 2006; Driskell et al. 2007; Hackett et al. 2008) and has suggested radically different phylogenies. Yet some of the strongest proponents of the use of phylogenetic analyses of the evolution of characters surprisingly maintained that they saw "no reason why our analyses should be systematically biased because they rest on unsatisfactory classifications" (Harvey et al. 1991)! I wonder if these authors would accept the sugges-tion that analyses of physiological rates would be equally acceptable if the rates were unsatisfactory. Any analysis of the evolution of character states depends critically on the accuracy of both character states and phylogeny. Symonds (1999) examined the impact of the structure of a phylogeny on analyses and concluded that the correlations found depend on the phylog-eny used. Besides, passerines do have higher mass-independent basal rates than the collective of other species ($P < 0.0001$) (see chap. 3 and fig. 3.4), although we can debate why this difference exists.

Many practitioners of phylogenetic analyses appear to be satisfied with a conclusion that a character state is the product of "history" without any attempt to clarify what aspect of history has produced the effect seen. Con-trary to the observation that basal rate in leaf-nosed bats of the Phyllosto-midae is *correlated* with food habits (McNab 1969), Cruz-Neto et al. (2001) concluded "that mass-independent BMR is not affected by diet, with simi-larities of species within clades being attributable to recency of common ancestry." This conclusion apparently means that frugivorous species have high basal rates because they evolved from other frugivorous species with high basal rates and that insectivorous species have low basal rates because they evolved from other insectivorous species with low basal rates. When does this determination end? And why do insectivorous phyllostomids have low basal rates and frugivorous species high basal rates? Is this correlation historically arbitrary and easily reversed? Besides, what is meant by "not *affected* by diet"? Determined by diet? The basal rates are surely *correlated* with diet (McNab 2003b), a statistical fact ($P < 0.0001$). The fundamental difficulty of separating the influence of phylogeny from those of ecology and behavior is that they are strongly correlated with each other.

Some recent analyses have examined the impact of phylogeny and the environment on the character states of species (Hansen 1997; Hansen & Orzack 2005; Hansen et al. 2008; Labra et al. 2009). In an analysis of the thermal biology of lizards belonging to the genus *Liolaemus*, Labra and colleagues concluded that "after controlling for adaptation, we found little evidence for phylogenetic inertia, and in most cases, a standard regression or ANOVA would have been a more accurate statistical procedure than an independent-contrast analysis." They added, "We suspect that a large number of published comparative studies have made unjustified corrections for phylogeny" (Labra et al. 2009).

The greatest failure of phylogenetic analyses occurs when quantitative characters are analyzed. A fundamental value of energy expenditure, besides its being a pivotal connector of an organism's performance with the environment, is its inherently quantitative nature, which permits a precise analysis of the contribution of factors to its numerical value. In an analysis of BMR in 639 species of mammals, 96.8% of the variation was accounted for by body mass alone (McNab 2008a). Even the strongest proponents of phylogenetic analyses cannot escape the quantitative dependency of BMR on body mass. The ability to account for the variation in mammalian BMR is significantly ($P < 0.0001$) increased to 97.9% by adding six other factors (see chap. 3). This increase in r^2 is modest because the range in mass is so great (i.e., it varies by a factor of 1.46×10^6:1), and r^2 approaches 100% as an asymptote. An ecologically diverse group of mammals with a smaller range in mass would give a clearer picture of the impact of factors other than mass. For example, measured phyllostomids have a mass range equal to 11.0:1. In this family, body mass alone accounts for only 78.7% of the variation in BMR, but the combination of body mass, food habits, distribution on continents or islands, and distribution at high or low altitudes accounts for 99.4% of the variation in BMR (see chap. 5). Of course, body mass would account for none of the variation in BMR in an ecologically diverse group of species of the same mass. The use of phylogenetic analyses does not address the numerical value of BMR and therefore yields a qualitative analysis of a quantitative relationship.

A quantitative analysis of BMR does not deny the impact of phylogeny, but simply attempts to account for the numerical impact of other factors, including body mass, on the BMR of endotherms. For example, if a species has a BMR equal to 20 kJ/h, why is it not 30 or 15 kJ/h, and what difference would it make if it were? This question is ignored by phylogenetic analyses, but is at the heart of a quantitative analysis. Besides, analysis of the energetics of birds and mammals includes the numerical impact of the passerine/nonpasserine and the eutherian/non-eutherian dichotomies (see chap. 3); that is, phylogenetic history obviously has an important impact on energy expenditure. *Attempting to account for numerical variations in energy expenditure is fundamental to understanding the relationship of a species to its environment, which is beyond what can be accomplished with a qualitative perspective.*

Species, irrespective of their ancestry, cannot expend energy at rates that are not ecologically justifiable given the resources they use, the environments in which they live, and the behaviors they engage in. *Energy expenditure is not determined willy-nilly.* It supports the body maintenance, activities, and fecundities of species. These concerns are most clearly seen in the numerical impact of various factors on energy expenditure. Even a phylogenetic factor, such as the reproductive dichotomy between marsupials and eutherians, has a quantitative impact (see chap. 13). The basis of a phylogenetic difference should be examined and not treated as a "black hole" that explains everything and nothing at the same time, as is the case when it is noted that marsupials have lower mean basal rates than eutherians because they are marsupials: circular reasoning.

Qualitative and quantitative approaches are asking different questions. One is concerned with the factors responsible for the level of quantitative characters—here, energy expenditure—whereas the other is interested in the historical origin of character states—that is, the historical *cause* for the occurrence of character states. The former depends on the quantitative consequences of *correlations* between character states, although the pursuit of correlations often provides insight into causation, at least of quantitative characters. These two approaches are not necessarily in conflict, as it sometimes appears. There should be room for both approaches. This book is principally occupied with analyzing the impact of factors that influence the level of energy expenditure in birds and mammals and the consequences of such correlations, although historical factors will repeatedly appear.

CHAPTER THREE. *A General Analysis of BMR*

The persistent preoccupation with equation 1.2, which ignores residual variation, suggests that little progress has occurred in the 80 years since the 1932 publication by Max Kleiber. This failure raises the possibility that factors other than mass influence BMR. Some of the first factors shown to affect BMR were climate (McNab & Morrison 1963) and food habits (McNab 1969, 1986a). Other factors potentially include habitat, substrate, torpor, the restriction of distribution, flight, and body composition, as well as affiliation with specific taxonomic groups (i.e., phylogeny). As we have seen, the principal difficulty with the attempt to identify the factors and their quantitative impact on energy expenditure is the association of these factors with evolutionary history. Data to analyze these potential relationships are available on nearly 1,200 species of birds and mammals, each of which is characterized by some eight to ten aspects of their life history; these data are summarized by McNab (2008a, 2009a).

Mammals

Seven factors in addition to body mass that potentially influence the BMRs of mammals have been examined. These factors include an island or continent and a lowland or montane distribution, each of which has 2 states, whereas climate has 5, food habits 22, habitat 7, substrate 7, and torpor 6, for a total of 129,360 possible combinations of states in the 639 species that have had their BMRs measured (McNab 2008a).

This complexity can be simplified in an ecologically and biologically relevant manner. Because the mean mass-independent BMRs associated with each state of a factor often are not statistically different from those associated with the other states of the same factor, statistically indistinct state categories can be combined. (Mass-independent characteristics are expressed as a decimal fraction of the value expected from mass.) Consider a simple case, thermal climate. Mammals were placed into 5 climate categories: polar, temperate, widespread, temperate/tropical, and tropical. The mean BMRs of polar ($P = 0.68$) and widespread ($P = 0.90$) species were indistinguishable from the mean BMR of temperate species, whereas the mean BMR of tropical/temperate species was different ($P \leq 0.0001$) from that of temperate species, but not different ($P = 0.91$) from the mean BMR of tropical species. Therefore, climate as a factor influencing the mass-independent BMR of mammals condensed to 2 categories, polar/temperate/widespread and tropical-temperate/tropical, the means of which were significantly ($P \leq 0.0001$) different. Similar condensation occurred in the other factors that had more than 2 categories: torpor condensed to 2 categories, substrate to 3, habitat to 4, and food habits to 3, each of which had a mean BMR that was significantly different from those of the other condensed categories of the same factor (McNab 2008a). Now the potential combinations of these factors decreased to 576, which is much more utilitarian.

As seen in chapter 1, $\log_{10}$ body mass alone accounted for 96.8% of the variation in $\log_{10}$ basal rate of metabolism in the 639 measured species (McNab 2008a). The seven condensed factors brought into the analysis were statistically significant individually and in combination. They and mass account for 97.9% of the variation in $\log_{10}$ BMR, what appears to be a small increase, but r^2 slowly approaches 100% as an asymptote.

A difficulty with this analysis remains: even after all these factors are included, the resulting equation overestimates the BMR of marsupials and monotremes; the mean measured BMR of marsupials is only 79.7% $\pm$ 1.98% ($n = 72$) of the value expected from this relationship, and it is only 61.8% $\pm$ 3.64% of the expected value for three monotremes. We know, for example, that eutherian grazers and carnivores have higher BMRs than marsupials with the same habits (McNab 1978b, 2005b, 2008a). Therefore, the differences in energy expenditure among eutherians, marsupials, and monotremes are not simply due to ecological or behavioral differences among these groups. So another term was introduced into the analysis: subclasses/infraclasses Monotremata, Metatheria, and Eutheria. It was a significant ($P \leq 0.0001$) correlate of $\log_{10}$ BMR when combined with $\log_{10}$ mass and the seven ecological/behavioral factors. That is, phylogeny itself had a significant impact on mammalian BMR (see chap. 11) independently of those factors.

Still, 97.9% of the variation in $\log_{10}$ BMR of mammals is accounted for, but when taken out of logs, 98.8% of the variation in BMR is accounted for by the following equation (McNab 2008a):

$$\text{BMR (kJ/h)} = 0.064 \, (M \cdot I \cdot S \cdot T \cdot C \cdot H \cdot E \cdot F) \, g^{0.694}, \qquad (3.1)$$

where eight dimensionless coefficients describe the response of BMR to various factors: M for mountains, I for islands, S for substrate, T for torpor,

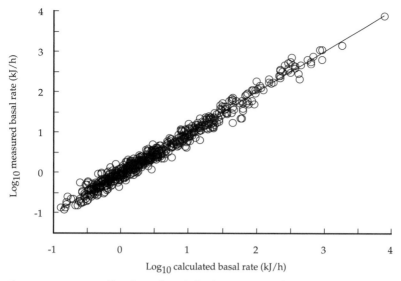

Figure 3.1. Log_{10} measured basal rate of metabolism in mammals as a function of log_{10} basal rate of metabolism calculated from equation (3.1). (Modified from McNab 2008a.)

C for climate, *H* for habitat, *E* for infraclass, and *F* for food habits. The estimated values for categories in these coefficients are summarized in table 3.1. Now the mean BMR of marsupials is 102.8% ± 2.47% ($n = 72$) of the value expected from equation (3.1), which is not different from 100%, and the mean BMR of the three monotremes is 100.2% ± 3.91%. Equation (3.1) reduces the residual variation from 3.2% to 1.2% (compare fig. 3.1 with fig. 1.5A).

A further complication exists: the analysis given above assumes that each of these factors acts independently of the others, which may not be the case. For example, it was unable to account for the variation of BMR in some ecologically defined groups. One such case was the persistent overestimation of BMR in frugivorous and folivorous arboreal carnivorans, including *Potos, Nandinia, Arctogalidia, Paradoxurus, Arctictis,* and *Ailurus* (mean = 58.8% of the value expected from eq. [3.1], $n = 6$) (fig. 3.2). Another was the underestimation of BMR in *Sorex* shrews (mean = 195%, $n = 7$) and arvicoline rodents (mean = 115%, $n = 33$) that live in temperate and polar climates and are dedicated to continuous endothermy (see chap. 4).

In each case, a suite of characteristics appears to be linked. Food habits, climate, and substrate define the arboreal frugivorous carnivorans. (It should be noted here that these species are called "carnivorans" because, although they are members of the order Carnivora, they are not carnivorous in the sense of feeding on vertebrates.) When this interactive factor was added to equation (3.1), it was a significant correlate of log_{10} BMR

TABLE 3.1. Dimensionless coefficients for the ecological characteristics of mammals

Factor	Categories	Value
M: Mountains/lowlands	Lowlands	0.88
	Mountains	1.00
I: Islands/continents	Islands	1.00
	Continents	1.21
T: Torpor	Torpor/hibernation	1.00
	No torpor	1.24
S: Substrate	Burrow/fossorial	0.70
	Trees/caves	0.81
	Terrestrial/aquatic	1.00
H: Habitat	Deserts	0.84
	Xeric/grasslands/savannahs	1.00
	Mesic/freshwater	1.19
	Marine	2.37
C: Climate	Tropical/temperate-tropical	1.00
	Temperate/polar/widespread	1.21
F: Food	Blood/ants-termites	0.71
	Seeds/leaves/insects/omnivory	0.90
	Vertebrates/fruit/nectar/grass	1.00
E: Infraclass	Monotremes	0.58
	Metatherians	0.84
	Eutherians	1.00

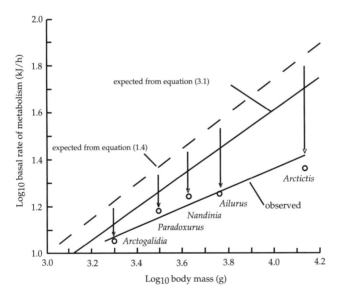

Figure 3.2. $\log_{10}$ basal rate of metabolism as a function of $\log_{10}$ body mass in arboreal carnivorans that feed on leaves or fruit, compared with the values expected from equations (1.4) and (3.1). (Modified from McNab 2008a.)

($P \leq 0.0001$), while all other factors retained a significant contribution to BMR, now accounting for 98.6% of its variation. However, when taken out of logs, this augmented analysis did not improve r^2, which remained equal to 0.988. But arboreal frugivorous carnivorans now had BMRs that averaged 106% $\pm$ 14%, although with great variation.

The second interaction, among climate, torpor, and body mass, was significant ($P = 0.010$) when added to equation (3.1), but lost significance when added with the frugivorous carnivorans to the equation. Therefore, the BMRs of soricine shrews and arvicoline rodents remain underestimated, which may mean that the apparent distinctness of soricines and arvicolines actually is not due to their shared commitment to a small mass, continuous endothermy, and a cold climate, or may related to some other factor, as appears to be the case in the marsupial/eutherian dichotomy. What might that factor be? In the case of the marsupial/eutherian dichotomy, it is related to the difference in the anatomical basis of reproduction and its consequence for energy expenditure (see chap. 12). However, the factors distinguishing soricine shrews and arvicoline rodents are unclear. Therefore, although appreciable progress has been made in accounting for the factors that determine BMR in mammals, we still have a way to go. In any case, this analysis is a significant improvement in the attempt to account for the variation in mammalian BMR.

Birds

A multifactor analysis similar to that just described for mammals was performed for 533 species of birds (McNab 2009a). The ecological/behavioral characters examined were food habits, thermal climate, habitat, altitude (restriction to mountains or lowlands), restriction to islands or continents, use of torpor, migration, and a volant or flightless condition. A restriction to islands or continents in birds was not significant when combined only with flight or migration and mass because of the sedentary or flightless state of many island endemics. Migration was not significant if only climate, altitude, and mass were included, principally because migration occurs in relation to climate and because most (studied) montane species are sedentary (as we will see shortly), but some of these combinations were statistically significant when other factors were included. The various factors condensed, as they did in the mammalian analysis: food habits condensed to 3 categories, climate to 3, habitat to 3, and torpor to 2, which, along with 2 conditions each in altitude and flight, led to 216 combinations, only 38% of the number found in mammals. Some of the acceptable states found in mammals, including a fossorial existence and sedentary arboreality, are incompatible with bird life, probably reflecting the influence of flight.

Even after six statistically acceptable ecological/behavioral factors were included in the analysis, a distinction in BMR between passerines and the collective of other orders was significant ($P \leq 0.0001$). Thus, the variation of BMR in birds is described by

$$\text{BMR (kJ/h)} = 0.139 \, (F \cdot A \cdot C \cdot E \cdot T \cdot U \cdot S) \, m^{0.689}, \qquad (3.2)$$

where the dimensionless coefficients F stand for food consumed, A for altitude, C for climate, E for avian order, T for torpor, U for habitat, and S for flight. The values for these coefficients are given in table 3.2. This relationship accounts for 97.7% of the variation in BMR (fig. 3.3A), which is clearly demonstrated even at the largest masses and highest BMRs (fig. 3.3B). This precision is greater than that found when only mass was included ($r^2 = 94.2\%$).

Notice that in the complete analysis represented by equations (3.1) and (3.2), unlike that represented by equations (1.4) and (1.5), the bird and mammal curves are parallel to each other; that is, they have essentially the same powers of mass: 0.694 ± 0.005 in mammals and 0.689 ± 0.008 in birds. The mean BMR of birds is greater than that of mammals of the same mass, roughly about $0.139/0.064 = 2.17$ times that found in mammals when a series of factors is considered in both classes. Why should birds have higher BMRs than mammals?

TABLE 3.2. Dimensionless coefficients for the ecological characteristics of birds

Factor	Categories	Value
F: Food habits	Fruit	0.69
	Insects/seeds/omnivory/vegetation/ aquatic invertebrates/vertebrates	0.74
	Nuts/nectar/pollen	1.00
A: Altitude	Lowlands	0.93
	Mountains	1.00
C: Climate	Tropical	1.00
	Temperate	1.16
	Polar	1.41
U: Habitat	Deserts/grasslands/savannahs/tundra	0.70
	Forests/disturbed/lakes/wetlands	0.80
	Marine/pelagic	1.00
T: Torpor	Torpor/?/hibernation	1.00
	No torpor	1.28
S: Flight	Flightless	1.00
	Flighted	1.35
E: Nonpasserine/ Passerine	Nonpasserines	0.76
	Passerines	1.00

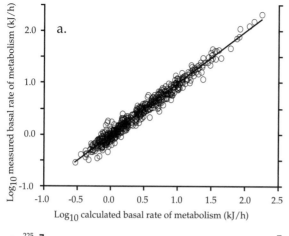

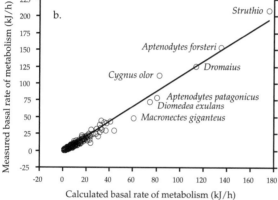

Figure 3.3. (A) Log_{10} measured basal rate of metabolism in birds as a function of log_{10} basal rate of metabolism calculated from equation (3.3). (B) Measured basal rate of metabolism in birds as a function of the basal rate of metabolism calculated from equation (3.3). (Modified from McNab 2009a.)

In the bird analysis, one of the factors influencing BMR is a volant or flightless condition. Volant species have basal rates that average 1.35 times those of flightless species. The 22 flightless species have basal rates that are only 105.3% of the values predicted by the eutherian curve. In contrast, eutherian fliers—bats—have basal rates that are indistinguishable ($P = 0.76$) from those of flightless mammals. The higher basal rates of birds appear to be principally associated with the morphology of avian flight, especially large pectoral muscle masses and possibly other features associated with flight, including large heart masses. The flight of bats is very different in structure from that of birds, in part because of their smaller flight muscle masses (Hartman 1961, 1963). Indeed, the cost of flapping flight in (some) small bats is 20%–25% less than in small birds (Winter and von Helversen 1998).

White et al. (2007a) examined the factors influencing the BMR of birds. Desert species, which face high temperatures and low rainfall, tend to have lower basal rates than more mesically distributed species. An explanation for this pattern might be that low basal rates are associated with low community production. An example of this association was found by Mueller and Diamond (2001) in small mammals. White and colleagues noted, however, that the basal rates of birds from the wet tropics were usually lower than those of temperate species, as did Wiersma et al. (2007), a pattern that does not correlate with community production. White et al. showed that BMR correlated with body mass, as expected, as well as with mean annual ambient temperature, temperature range, and precipitation. They concluded that the basal rate of birds was better accounted for by including all three of those factors than by including only one in addition to body mass. They also concluded that it was difficult to understand why basal rates were correlated with precipitation in birds, whereas they were not in mammals, but the animals in their "mammals" category were actually marsupials, which in many ways (see chap. 13) are not generally representative of mammals.

As we have seen, appreciable variation in basal rate in both mammals and birds is related to a variety of ecological and behavioral factors; these patterns will be explored in chapters 4 through 8. A question of interest, however, is how these biological factors exert their effects on BMR. One means is through their association with body composition.

Body composition

The total metabolic rate of an organism is the sum of the metabolic rates of each of its tissues and organs. This was clearly shown by Wang et al. (2001), who tried to account for Kleiber's (1932) scaling relationship for mammalian BMR in domesticated mammals,

$$BMR \ (kcal/d) = 70 \ kg^{0.75},$$

by summing the rates of metabolism of various tissues and organs. They demonstrated that some tissues and organs had high mass-specific rates of metabolism, including heart, kidneys, liver, and brain. When the rates of the tissues were summed, Wang and colleagues constructed the relationship

$$BMR \ (kcal/d) = 67 \ kg^{0.76},$$

which is similar to that of Kleiber and accounts for the scaling of metabolism by a power b that is <1.00. Body composition, obviously, is a very important factor determining the level and scaling of energy expenditure. However, this attempt to estimate the total rate by summing the rates of individual tissues neglects the cost of their integration, including the cost of thermoregulation in endotherms.

Evidence has accumulated that body composition influences some of the residual variation in BMR. This might be the case because not all of the tissues that make up a species' mass are equally metabolically active. Fatty tissue, for example, has a low metabolic activity. So an individual or species with large fat deposits might have a lower total rate of metabolism than a lean individual of the same total mass. Or, a species with a large amount of tissue having a high rate of metabolism, as has been suggested for brain tissue, might have a high BMR for its mass (Martin 1981, 1998).

Several examples have indicated a correlation of BMR with body composition. Shorebirds have basal rates that correlate with their heart and kidney masses (Daan et al. 1990). The seasonal variation in BMR in the redshank (*Tringa totanus*) is associated with changes in both lean and fat masses (Scott et al. 1996), but BMR is not correlated with seasonal variation in the size of fat deposits in northern phocid seals (Ochoa-Acuña et al. 2009). Rails have basal rates that correlate with their food habits and body composition; herbivorous gallinules have higher basal rates and larger pectoral muscle masses than omnivorous rails (McNab & Ellis 2006). Mammals that are committed to arboreal lifestyles, including a tree kangaroo (McNab 1988a), tree sloths (McNab 1978c), various viverrids (McNab 1995), and the margay (*Leopardus wiedii*; McNab 2000c), have lower basal rates than their terrestrial relatives, possibly because of reduced muscle masses.

An example of the association of body composition with energy expenditure is found in raptors: species that pursue prey in flight, especially those belonging to the genera *Accipiter* and *Falco*, have higher mass-independent basal rates (97.0% [$N = 3$] and 95.5% [$N = 2$], respectively) than species that search for prey either from a perch or by soaring—namely, species that belong to the genera *Buteo* (72.0% [$N = 2$]), *Parabuteo* (56%), *Aquila* (70%), or *Ictinia* (65%) (Wasser 1986). This difference correlates with the size of the pectoral muscle and heart masses: they are larger in *Accipiter* and *Falco* (13%–22% and 0.73%–1.2% of total mass) than in *Buteo* (12%–14% and 0.4%–0.6%) (Hartman 1961). An exception is found in *Buteo buteo*, which apparently has a high BMR, 102% (Prinzinger & Hänssler 1980), for an unclear reason. *Falco sparverius*, a hovering, insectivorous falcon, is intermediate in having a pectoral muscle mass equal to 15% and a low BMR (74%). Of special interest is the osprey (*Pandion haliaetus*), which searches for prey (fish) by soaring, but must make an appreciable effort to lift itself and a fish, which may weigh up to 1 kg, out of water, so it is not surprising that it has a high BMR (117%), even though its pectoral muscle and heart masses are intermediate (14.7% and 0.84%, respectively). The New World black vulture (*Coragyps atratus*) and the turkey vulture (*Cathartes aura*) have intermediate to large pectoral (14.3% and 16.4%, respectively) and heart masses (0.85% and 0.66%) (Hartman 1961). The black vulture has a higher wing loading than the turkey vulture (0.64 and 0.33 g/cm^2) and consequently spends more time using flapping flight. Unfortunately, no ca-thartid has had its basal rate measured.

Basal rates in nonpasserine birds that search for insect prey with sus-tained flight, such as swifts (68.3% $\pm$ 10.87% [$N = 3$]) and the nighthawk (*Chordeiles minor*, 67%), a continuously active caprimulgid, are higher than in related species that sit and wait for prey, including caprimulgids that be-long to the genera *Phalaenoptilus, Caprimulgus,* and *Eurostopodus* (48.0% $\pm$ 3.08% [$N = 4$]) as well as frogmouths (Podargidae) (56.7% $\pm$ 2.73% [$N = 3$]) (McNab & Bonaccorso 1995). The lower basal rates in the sit-and-wait species may be related to smaller pectoral muscle masses, but few species have been examined (Hartman 1961). Insectivorous bats that are sit-and-wait predators have lower basal rates than those that persistently pursue insects (Bonaccorso & McNab 2003). Thus, the impact of food habits on the basal rate in endotherms that fly may depend on the means by which they access their food supply and their body composition. This pattern appar-ently does not occur in passerines (Bonaccorso & McNab 2003).

Whether passerines have higher BMRs than other birds has been the subject of extensive controversy (see chaps. 2 and 5). If a difference in BMR between passerines and other avian orders actually exists, why should it be present? One of the difficulties in answering this question is that the majority of the passerines measured live in the temperate zone, and those species have higher BMRs than tropical passerines (Wiersma et al. 2007; McNab 2009a). The higher basal rates of temperate passerines might re-flect a climatic adaptation. However, an analysis of the BMRs of 533 species of birds (McNab 2009a) showed not only that passerines have high mass-independent basal rates (fig. 3.4; see eq. [3.2]), which averaged 132% of the collective of other birds ($P < 0.0001$), but that four nonpasserine orders have BMRs that are indistinguishable from those of passerines. This is not surprising, given the great ecological diversity among the orders.

The orders that do not differ from passerines ($n = 272$) are Anseriformes ($P = 0.20, N = 30$), Charadriiformes ($P = 0.81, N = 25$), Procellariiformes ($P = 0.26, N = 13$), and Pelecaniformes ($P = 1.00, N = 4$). What these four orders share is a high degree of mobility associated with widespread occurrence in temperate and polar regions, characteristics that they share with many passerines. Their high level of activity may be associated with a body composition that facilitates migration (e.g., large pectoral muscles with high mitochondrial densities, a highly efficient circulatory system, and a large heart). Gavrilov (1995, 1998, 1999) was the first to suggest that the high basal rate of temperate passerines is related to migratory distance. A corollary of this analysis is that sedentary species belonging to these four

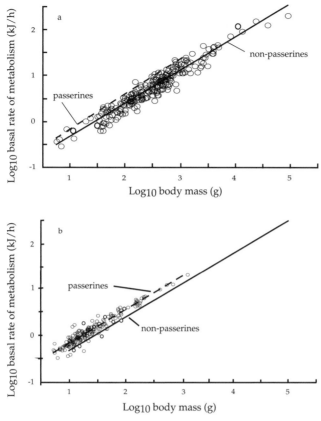

Figure 3.4. Log$_{10}$ basal rate of metabolism as a function of log$_{10}$ body mass in (A) frugivorous nonpasserines and (B) frugivorous passerines. (Data from McNab 2009a.)

nonpasserine orders would probably have low BMRs. In fact, the sedentary flighted and flightless anseriforms endemic to temperate New Zealand have mass-independent BMRs that are 70% of those of the highly migratory north-temperate anseriforms (McNab 2003a), and the lower mass-independent BMRs in these species correlate with the smaller size of their pectoral muscle masses (see chap. 9).

The difficulty with these analyses is that they are at best correlative and thus not necessarily indicative of causative factors. For example, Martin (1981, 1998) suggested that brains are costly and therefore species with large brains should have high BMRs, although Wang et al. (2001) found that the mass-specific metabolic rate of brain tissue was lower than that of the liver, kidneys, and heart, but much higher than that of the "remaining tissues." McNab and Eisenberg (1989) concluded that there was no correlation ($P > 0.05$) of mass-independent BMR with mass-independent brain size in 174 species of mammals. (Note that brain mass and BMR both correlate with body mass, so the variation in both characters must be examined independently of body mass; otherwise, a correlation between them may reflect their mutual correlation with mass.) Another analysis of brain size and BMR in 49 species of bats (McNab and Köhler, unpublished) also demonstrated no correlation ($P = 0.53$) of brain size with BMR.

Determining whether particular tissues have high energy expenditures by measuring the total rate of metabolism presents many problems. As the proportion of a particular tissue with high metabolic activity increases, that of other tissues with high activity may decrease, leaving the total rate of metabolism unchanged. Only if low-activity tissues are proportionally replaced by high-activity tissues would the addition be reflected by an increase in total metabolism. These trade-offs are always the problem with correlations: rarely does only one factor change at a time, so the establishment of cause and effect is usually difficult.

This concern with correlations will persist throughout this text because the analysis of most physiological/ecological relationships is fundamentally based on correlations, and therefore all analyses are at best suggestions. But suggestions are better than no analysis, such as assuming that everything is due to phylogeny without any understanding of the functional bases of changes in character states. Most suggestions can be further evaluated by examining their potential consequences, which may reinforce or refute the suggested conclusions. This is how intellectual progress is usually accomplished.

A general conclusion that appears appropriate is that scaling relationships reflect "engineering," whereas the residual variation in those relationships reflects the impacts of ecology and behavior. Consequently, much more attention should be paid to residual variation in the attempt to understand the problems faced by species and the solutions that they have evolved. The next chapters of this book will be concerned with the consequences of residual variation.

CHAPTER FOUR. *Small and Large*

The dramatic influence of body size on many aspects of the life histories and energy expenditures of endotherms is most clearly seen at the smallest and the largest masses.

Life at the smallest masses

Endotherms have a limited number of responses to small size. They include a limitation to a small mass, the use of daily torpor, and a commitment to continuous endothermy.

Limitation to a minimal size

The smallest adult birds and eutherian mammals weigh between 1.6 and 2.5 g. In eutherians, very small masses are found in shrews, both soricine (e.g., *Sorex hoyi* at 2.1 g and *S. minutissimus* at 2.5 g) and crocidurine (e.g., *Suncus etruscus* at 2.2 g), and in the bumblebee bat (*Crasseonycteris thonglongyai*, ca. 1.7 g) from Thailand. Some bats in the Hipposideridae (e.g., *Ascelliscus*) and Emballonuridae (e.g., *Saccopteryx*) weigh as little as 4 g. The smallest marsupial is the long-tailed planigale (*Planigale ingrami*), which weighs 4–6 g. The smallest birds are hummingbirds, many of which have masses between 2.5 and 6.5 g; the smallest species is the bee hummingbird (*Mellisuga helenae*) from Cuba, at 1.6–1.9 g, whereas the largest is the "giant" hummingbird (*Patagona gigas*) from Andean South America, at 18.5–20.5 g. Some passerines weigh <6 g, including kinglets (Regulidae), tits (Aegithelidae), gnatcatchers (Polioptilidae), and sunbirds (Nectariniidae).

A decrease in mass reduces energy expenditure. This statement is derived from equation (1.2) and the realization that it is the total, not the mass-specific, rate of metabolism that is ecologically significant. (The only species that live on a per gram basis weigh 1 gram: all species live on the basis of their total masses, which is why total rates of metabolism are used throughout this book.) One of the most distinctive responses to cold temperatures by some small mammals is an autumnal reduction of mass in anticipation of a decrease in ambient temperature and a potential shortage of food, a phenomenon first described by August Dehnel (1949). This decrease in mass, which is called Dehnel's phenomenon, leads to a reduction in energy expenditure (McNab 2010), a stratagem that is explored in greater detail in chapter 6.

A reduction in the mass of an endotherm, however, cannot occur without consequences. One is that a mass will be reached below which a dichotomy must be addressed: either the animal continues the reduction in mass without any adjustment in the scaling of metabolism, thereby forfeiting continuous endothermy with an entrance into torpor, or it adjusts the scaling of metabolism so as to maintain a commitment to continuous endothermy. This dichotomy is addressed at masses <45 g.

The use of torpor

Torpor has at least two forms; one is used opportunistically on a daily basis, at night in diurnal species and during the day in nocturnal species, and the other is used for extended periods: aestivation in summer and hibernation in winter. Seasonal torpor is best viewed as a response to harsh environmental conditions, a seasonal shortage of resources, and an inability to evade these conditions (see chap. 8), whereas daily torpor use by small species is predominantly a response of an endotherm with a small mass to a short-term shortage of resources.

A common response to a very small mass, as we have seen in endothermic insects, is intermittent endothermy. This is the response seen in crocidurine shrews, vespertilionid bats, hummingbirds, all small dasyurid marsupials, and undoubtedly in *Crasseonycteris*. It also occurs in many small rodents, especially in members of the Cricetidae: the smaller they are, the more likely they are to enter torpor. Thus, within the subfamily Neotominae, the small *Baiomys taylori* (7.3 g) readily goes into torpor, as does *Reithrodontomys megalotis* (7.6 g), but *Peromyscus* species (12–50 g) generally go into torpor with some reluctance, especially the larger species, and the much larger (140–320 g) wood rats (*Neotoma*) have never shown evidence of torpor. Among the Heteromyidae, pocket mice (Perognathinae) of both small (8 g) and large (40 g) species use torpor, most notably in arid environments (see chap. 7). The only kangaroo rat known to enter torpor is *Dipodomys merriami* (Carpenter 1966), one of the smallest species (38 g); larger species (49–106 g) have not shown this behavior.

The occurrence of daily torpor is much less common in birds than in mammals, principally because birds have the capacity to avoid seasonally harsh conditions by migration. They generally have much higher basal rates than similarly sized mammals, which allow them to avoid going into torpor (as we will see shortly). Furthermore, their high basal rates contribute to the lower scaling power *b* in the avian BMR-scaling curve (see chap. 1). Torpor in birds, when it occurs, is primarily a response to a small mass and thus occurs in tropical as well as temperate environments. Among birds, daily torpor is best known in hummingbirds, most of which are tropical. Female Puerto Rican todies (*Todus mexicanus*), which weigh only 5–7 g, enter torpor during the breeding season (Merola-Zwartjes & Ligon 2000). Some sunbirds (Nectariniidae) also enter torpor, including *Aethopyga christinae* (5.5 g) and *Nectarinia famosa* (15–20 g) (Prinzinger et al. 1992; Downs & Brown 2002). Other small tropical birds may use daily torpor, but few have been studied from this viewpoint.

A commitment to continuous endothermy

Unlike the two responses to small size that decrease energy expenditure, reduction in mass and the use of torpor, continuous endothermy at a small mass markedly increases energy expenditure relative to the standard metabolism-mass curve (McNab 1983).

Some very small mammals show no evidence of using torpor (McNab 1983). This pattern is most marked in shrews of the genus *Sorex*, members of the subfamily Soricinae (McNab 2006a). They are apparently continuously endothermic, which at a mass as small as 2.1 g would require them to have exceedingly high mass-independent basal rates of metabolism (fig. 4.1). For example, *Sorex minutus*, at 3.3 g, has a basal rate of metabolism that is 3.4 times the rate expected from the general mammalian metabolism-mass curve (Sparti & Genoud 1989). In contrast, nearly all studied species in the Crocidurinae, another subfamily of the Soricidae, which includes *Crocidura* and *Suncus*, enter torpor, except possibly *C. oliveri*, and have lower basal rates than similarly sized *Sorex* (see fig. 4.1).

Suncus etruscus has a basal rate that is only 1.13 times the value expected from the standard curve for its mass of 2.2 g (Frey 1980). This multiplier is only one-third of that found in *Sorex minutus*, a difference undoubtedly reflecting the difference between a continuous endotherm (*Sorex*) and one that uses torpor (*Suncus*). If, however, *Sorex minutus* weighed 2.2 g and maintained a commitment to continuous endothermy, it would probably have a BMR as high as 3.9 times the standard rate. These observations make the basal rate in very small *Sorex* a subject of great interest, especially that of *S. hoyi*, which is found in Alaska and Canada and is the smallest mammal (at 2.1 g) in North America.

The closest New World physiological equivalent of *Crocidura* is the

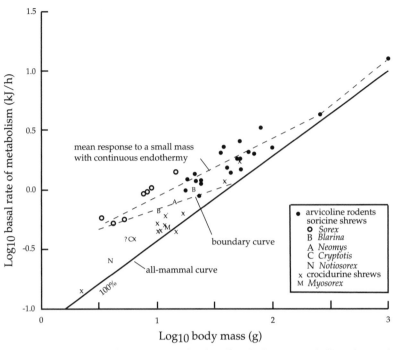

Figure 4.1. Log$_{10}$ basal rate of metabolism as a function of log$_{10}$ body mass in arvicoline rodents and shrews. (The boundary curve is derived from McNab 1983.)

soricine desert shrew (*Notiosorex crawfordi*), which is found in the Sono-ran Desert and, unlike all other studied soricines, uses torpor (Lindstedt 1980). It has a basal rate that is 137% of the value expected from the general mammalian curve. Most of the other soricines have basal rates from 167% to 345%, depending on body mass, although *Notiosorex* is somewhat larger (4.0 g), which may contribute, along with its desert distribution, to its somewhat lower mass-independent BMR. *Sorex preblei* would be worthy of study, not only because of it small mass (2–4 g), but because it lives the Great Basin desert in the western United States, which raises the possibility that it, like *N. crawfordi*, enters torpor, even though it is a *Sorex*. Other soricine genera, including *Blarina* and *Neomys*, have basal rates that are similar to those of *Sorex* shrews (see fig. 4.1). Adjusted for the difference in mass, measurements on *Cryptotis parva*, another soricine shrew, indicated a lower basal rate (146% of the standard curve), but it gave no evidence of entrance into torpor (McNab 1991). Diversity in BMR and in the selective occurrence of torpor in the Soricinae fundamentally undercuts the concept that much of physiological, behavioral, and ecological diversity simply reflects phylogenetic lineages.

Some Australian physiologists (White & Seymour 2003, 2004) have dismissed the very high basal rates reported for *Sorex* as being artifacts, but that appears to reflect their experience with small dasyurid marsupials, which enter torpor and have low basal rates. An extended attempt to explore the possibility that the high rates in soricine shrews were artifactual (McNab 1991) concluded that those measurements were reasonable estimates of BMR, as did a similar study of *Blarina brevicauda* (Dawson & Olson 1987).

Like soricine shrews, arvicoline rodents have failed to show evidence of entering torpor (see box 4.1), in part because they have a mass >15 g and in part because they have basal rates similar to those found in similarly sized *Sorex* (see fig. 4.1). Arvicoline rodents belong to a subfamily of the Cricetidae that includes *Arvicola, Microtus, Myodes* (= *Clethrionomys*), *Lemmus, Phenacomys, Arborimus,* and *Dicrostonyx,* among others. These rodents have high basal rates (McNab 1992b), but given their masses, which are greater than those found in *Sorex,* arvicoline rates are less exaggerated. The smallest arvicoline rodent measured, *Microtus subterraneus* (17.8 g), has a BMR that is 1.8 times that expected from the standard curve (Gebczynski & Szuma 1993). What is so interesting about the arvicolines is that the BMRs of species with masses >100 g scale higher than, but parallel to, the standard curve (see fig. 4.1), whereas smaller species have shifted the scaling of BMR to coincide with those of *Sorex,* thereby defining a mean response in small species committed to continuous endothermy.

These two groups, soricine shrews and arvicoline rodents, define a curve at masses <45 g that appears to be a boundary between mammals that are committed to continuous endothermy and those with lower basal rates that use daily torpor (see fig. 4.1). This mass-BMR relationship has been called the "boundary curve for endothermy":

$$\text{BMR (kJ/h)} = 0.31 \; g^{0.33} \tag{4.1}$$

(McNab 1983). At 45 g, a species conforming to this relationship would expend energy at 1.09 kJ/h, as it would conforming to the standard mammalian curve. At 10 g, a species would expend energy at 0.66 kJ/h if it conformed to the boundary curve and at 0.36 kJ/h if it conformed to the standard curve, now 1.8 times the standard value; at 2 g, this ratio would equal 3.4:1! It should be noted, however, that the influence of the boundary curve may be directional: *all* species that fall below the curve may enter torpor, but *some* of the species that fall above the curve also do so. As noted, *Notiosorex* uses torpor, and its basal rate falls well below the boundary curve, as does that of *Cryptomys,* which has not (yet?) shown evidence of torpor. *Neomys* and *Blarina,* however, either are on or above the boundary curve (see fig. 4.1).

An extreme example of the impact of this relationship might be found

BOX 4.1. ·

Rigid Endothermy in Arvicoline Rodents

In Colorado, my youngest son Derrick and I trapped mountain meadows at elevations of about 3,000 m during a period of cool, wet weather. Although I tried to check the traps often, occasionally I would pull a wet, cold vole out of a trap. I would dry the animal and warm it up, and it would appear to recover, but it always died the next day, which led me to conclude that cold-temperate arvicolids, and by implication, all arvicolids, are "rigid" endotherms incapable of tolerating a reduced body temperature. They are therefore very different from *Peromyscus*, which does not enter deep torpor or hibernation, but readily tolerates and easily recovers from a low body temperature.

· ·

in a 100 mg insect. If committed to continuous endothermy, it would have to have a BMR, estimated from the boundary curve, that was 10.9 times the standard mammalian rate, an absolute rate that is appropriate for a 2.7 g mammal conforming to the standard curve. Thus, the same amount of energy would be required to support a 0.1 g endotherm in continuous endothermy as to support a standard endotherm with 27 times the mass—a relationship that surely sets a lower limit to the mass at which continuous endothermy can be supported. Obviously, a commitment to continuous endothermy at small masses is exceedingly expensive, which explains why no insects make this commitment and why most of the smallest mammals and birds enter torpor: no species that weighs <2 g is known to be a continuous endotherm, and even then, few species marginally above that size are continuous endotherms.

Mammals that weigh >100 g and have high basal rates (i.e., >100% of the values expected from mass in the standard curve), including tree squirrels (fig. 4.2), lagomorphs (fig. 4.2), canids (fig. 4.2), arvicoline rodents (fig. 4.3), and murid rodents (fig. 4.4), scale parallel to the standard curve. In contrast, mammals that weigh <100 g and are characterized by basal rates <100%, such as fossorial mammals, scale in a manner that tends to coalesce with the high-BMR groups, with a few exceptions (fig. 4.2), thereby avoiding entrance into torpor.

However, small species that have a BMR that falls below about 1 kJ/h ($\log_{10} 1 = 0$), such as the marsupial mole (*Notoryctes caurinus*), small golden moles (Chrysochloridae), and *Heterocephalus*, enter torpor, but species whose BMR remains above that value, such as the chrysochlorid *Amblysomus*, talpids, and most small fossorial rodents, avoid the use of torpor. A similar pattern in heteromyids indicates that all species that have BMRs below 1 kJ/h (i.e., species belonging to the genera *Perognathus*, *Chaetodipus*,

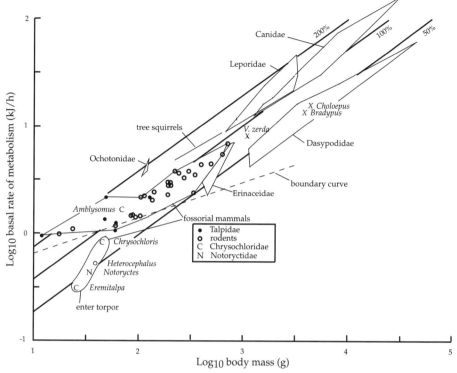

Figure 4.2. Log$_{10}$ basal rate of metabolism as a function of log$_{10}$ body mass in fossorial mammals and various eutherians. The standard curve for mammals, as well as curves that are twice and one-half of the standard curve, and the boundary curve are indicated, as are the fossorial mammals that enter torpor. (Data from McNab 2008a.)

and *Microdipodops*) enter torpor, as does *D. merriami*, the smallest kangaroo rat (fig. 4.3), whereas larger kangaroo rats and *Heteromys* do not enter torpor. Another small kangaroo rat, *D. nitratoides*, has a BMR similar to *D. merriami*, but has not been reported to enter torpor, and it is unclear whether *Liomys*, which is at the border between continuous endothermy and the use of torpor, enters torpor.

Murid rodents present a more complicated picture. Quite a few species have had their basal rates measured (fig. 4.4), especially species belonging to the gerbilline and murine subfamilies. The BMRs of gerbils scale in parallel to the all-mammal curve without any tendency to shift at small masses to conform to the boundary curve. Therefore, we should expect that small gerbils would enter torpor. However, the smallest species (*Gerbillus pusillus*), which has a basal rate above the all-mammal curve but well below the boundary curve, reduces BMR with a restriction in food availability (Merkt

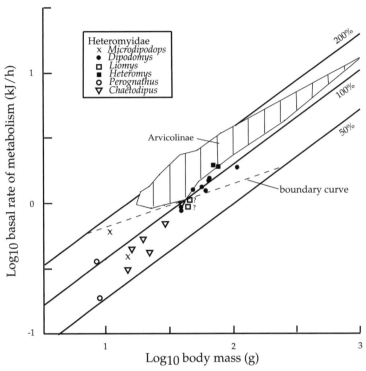

Figure 4.3. $\log_{10}$ basal rate of metabolism as a function of $\log_{10}$ body mass in heteromyid rodents, along with the mammalian standard curve at various levels and the boundary curve. Question marks indicate uncertainty whether some species go into torpor. A polygon indicates the basal rates of arvicoline rodents, none of which enter torpor. (Data from McNab 2008a.)

& Taylor 1994). This behavior was observed at an ambient temperature of 31°C, which would not facilitate an entrance into torpor. This species may well enter torpor with food restriction at low ambient temperatures.

The BMRs of murine rodents show a tendency to shift toward the boundary curve (see fig. 4.4), but generally fall below it at masses <40 g. The murine rodent *Pseudomys hermannsburgensis*, an Australian desert species, may use torpor (Predavec 1997), but some question has arisen whether the reduction in its metabolic rate actually represents forced hypothermia (Tomlinson et al. 2007). Downs and Perrin (1996) concluded that one of the smallest murines, *Mus minutoides* (8.3 g), does not enter torpor, although Webb and Skinner (1995) found low body temperatures in this species at an ambient temperature of 8°C. One deomyine, *Acomys russatus*, enters torpor (Ehrhardt et al. 2005) and has a mass-independent basal rate that falls well below the boundary curve (see fig. 4.4). Clearly, murines complicate the concept of the boundary curve, and appreciably more work needs to be done on desert species belonging to this family. (It should be noted that the demonstration of torpor in a species is not particularly easy, especially if entrance into torpor requires a trigger, which may not be given to captive individuals. Furthermore, the difference between marginal temperature regulation and entrance into torpor is difficult to distinguish.)

The boundary curve separates most marsupials that enter torpor from those that do not (fig. 4.5): all species found below that curve enter torpor (including hibernation in some species), as do four species on, or slightly above, the curve. The low basal rates of the small dasyurids conform to the general absence of high basal rates in marsupials, and at these small masses, reflect their propensity for torpor (see chap. 13). Neotropical marsupials belonging to the tribe Marmosini, which contains the smallest didelphid marsupials, are somewhat equivalent to dasyurids, but they have a minimal mass that is only 10–15 g. These mixed-diet marsupials undoubtedly enter torpor, but their energetics has been collectively ignored, except for a limited examination of *Gracilinanus* (= *Marmosa*) *microtarsus* from Brazil (Morrison & McNab 1962), which weighs 13 g. It has a basal rate that is 85% of the value expected from mass in mammals and enters torpor (see fig. 4.5). Another didelphid, *Thylamys elegans*, which weighs 40 g, readily goes into torpor and hibernation and has a basal rate that is 86% of the all-

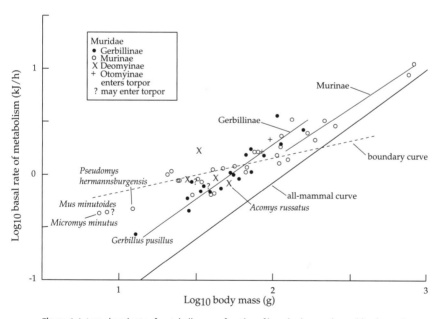

Figure 4.4. Log_{10} basal rate of metabolism as a function of log_{10} body mass in murid rodents. Curves for murines and for gerbillines are indicated, as well as the all-mammal standard, one-half the all-mammal standard, and boundary curves. Species that are know to enter torpor are indicated. (Data from McNab 2008a.)

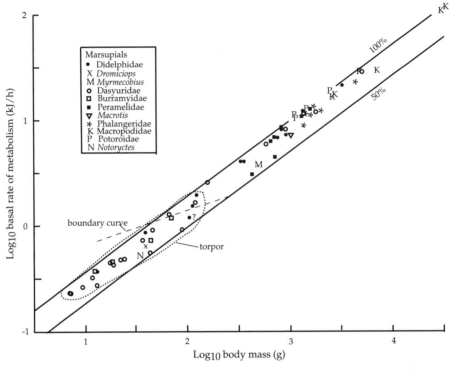

Figure 4.5. $\log_{10}$ basal rate of metabolism as a function of $\log_{10}$ body mass in marsupials. The all-mammal standard, one-half of the all-mammal standard, and boundary curves are indicated, as are the species that enter torpor. (Modified from McNab 2005b.)

mammal value (Bozinovic et al. 2005). However, *Monodelphis domesticus*, which marginally falls below the curve and weighs 104 g, has not been observed to enter torpor.

What is so astounding about the marsupial data is the narrow range in mass-independent basal rates, with none high and only a few as low as 50% of the all-mammal curve (see fig. 4.5). This twofold range in marsupials contrasts with that of eutherians, which have at least a twelvefold (0.3–3.8) range in mass-independent basal rates (see chap. 13). No small marsupials increase BMR to avoid entrance into torpor, as fossorial insectivores and rodents, and even elephant shrews, do. The restricted response of marsupials to internal and external factors is explored in chapter 13.

A critical examination of the boundary curve by Cooper and Geiser (2008) found that it distinguished correctly between species committed to continuous endothermy and those with an endothermy that involves the use of daily torpor in 82% of rodents, 88% of marsupials, and 94% of bats, which led these authors to conclude that this relationship is inaccurate. The extent to which the boundary curve "errs" in predicting the performance of individual species may reflect its failure to account for factors other than mass that modify this relationship, which in its simplest form is remarkably accurate. When dealing with a character as complex as energy expenditure and its consequences, as we have seen in chapter 3, no biological "rules" are simple or absolute, because species have the ability to circumvent apparent limitations that has permitted the diversity of species to evolve on this planet. To account for these secondary factors, more complex forms of these "rules" are required.

The variable scaling of BMR points out that a trade-off can potentially occur between the rate of energy expenditure and body size: if total energy expenditure is limited by resources, the limited energy available can be apportioned either to maintenance or to growth (fig. 4.6). For example, terrestrial insectivorous mammals whose BMRs equal 0.50 kJ/h ($\log_{10} = -0.30$) occur over a range in mass from 3.7 to 25.0 g—namely, from soricine shrews to crocidurine shrews and small dasyurids. The shift to larger masses reflects

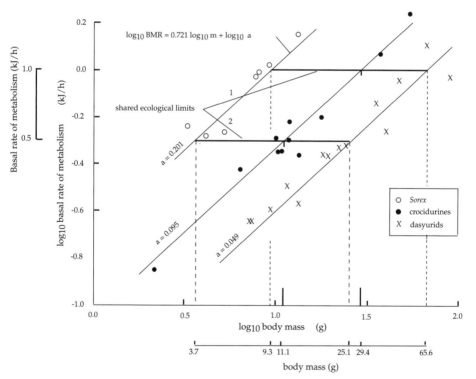

Figure 4.6. $\log_{10}$ basal rate of metabolism as a function of $\log_{10}$ body mass in *Sorex* and crocidurine shrews and dasyurid marsupials. Also included are two arbitrary levels of energy expenditure, 0.5 and 1.0 kJ/h. (Data from McNab 2008a.)

a progressive reduction in the level of their mass-independent BMRs, as reflected in the decrease in the coefficient *a* in equation (1.2). In other words, insectivorous mammals can be smaller or larger at the same level of energy expenditure; the smaller species have mass-independent energy expenditures high enough to be continuously endothermic, whereas crocidurines and dasyurids at the same energy expenditure enter torpor. If the resources used by insectivorous mammals permit a BMR equal to 1.00 kJ/h ($\log_{10} = 0.0$), body mass ranges between 11.5 and 66.8 g, and soricine and crocidurine shrews and dasyurids do not have to enter torpor. As a consequence, large shrews are found in rich habitat patches and small shrews in barren patches (Hanski & Kaikusalo 1989).

If the boundary curve for endothermy in mammals actually describes a fundamental relationship between energy expenditure and body mass in endotherms in the absence of torpor, it should also apply to small birds that do not enter torpor. Indeed, the BMRs of most birds fall into the polygon that contains all continuously endothermic mammals (fig. 4.7). For example,

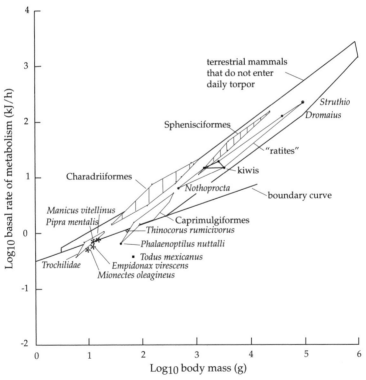

Figure 4.7. $\log_{10}$ basal rate of metabolism as a function of $\log_{10}$ body mass in various nonpasserine birds, shown in relation to a polygon for terrestrial mammals that do not enter torpor. (Data from McNab 2008a, 2009a.)

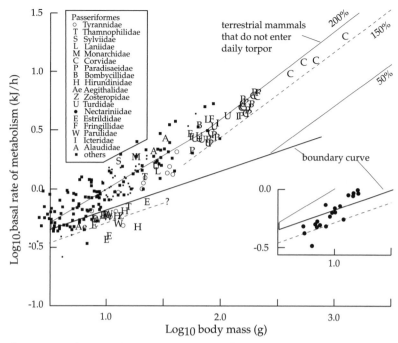

Figure 4.8. Log$_{10}$ basal rate of metabolism as a function of log$_{10}$ body mass in passerines, along with the mammalian standard curve at various levels and the boundary curve. The inset graph indicates the basal rates of some insectivorous and frugivorous birds that weigh <12 g and fall on, or below, the boundary curve, and a dashed curve that may be a more appropriate boundary curve for passerines. (Data from McNab 2009a.)

penguins, ratites, caprimulgids, and most charadriiforms and some tro-chilids fall within the mammalian polygon, whereas some of the shorebirds fall above it. However, other hummingbirds, the poorwill (*Phalaenoptilus nuttallii*), and the Puerto Rican tody, which enter torpor, fall below the boundary curve (fig. 4.7).

A close examination of passerines (fig. 4.8) indicates that the vast major-ity of species fall into the mammalian polygon. The basal rates of passerines that weigh >100 g generally fall between 150% and 200% of the values ex-pected from the mean mammalian curve, especially those of corvids, birds of paradise, icterids, and thrushes. Most passerines with masses <30 g have basal rates similar to those found in *Sorex* shrews and arvicolid rodents, re-flecting the tendency of temperate and polar birds to avoid torpor, depend-ing instead on their food habits and their ability to avoid winter's harshness through migration. Birds, therefore, have an environmental escape clause (see chap. 8) that is denied to small mammals, except for a few bats. A few small passerines have basal rates that are greater than 200% of the mammalian curve, including a sylviid (*Hippolais icterina*), two larks (*Lullula arborea* and *Alauda arvensis*), a waxwing (*Bombacilla garrulus*), and a shrike (*Lanius excubitor*), all of which are cool- to cold-climate species. The observation that most passerines have higher basal rates than other birds might be related to, or simply result in, their maintenance of continuous endothermy at small masses, which is why their BMRs coincide with those of soricine shrews and arvicoline rodents.

Some passerines fall below the boundary curve, especially swallows (Hirundinidae) and sunbirds (Nectariniidae), both of which have been shown to include some species that enter torpor. Other species that belong to families that have not been shown to enter torpor also fall below the curve, including warblers (Parulinae), finches (Estrilidae), and flycatchers (Tyrannidae). These observations raise two possibilities: (1) some of these species will be shown to enter torpor, or (2) the boundary curve for birds might be set at a lower level than in mammals; that is, at $a = 0.24$ (see dashed curve in fig. 4.8), rather than 0.31, in equation (4.1).

Finally, the boundary curve even applies to ectotherms generally, at least at masses <10 kg (fig. 4.9). The skipjack tuna (*Katsuwonus pelamis*), which regulates lateral muscle temperatures, is in the endothermic group, and the boundary curve separates incubating pythons (*Python molurus* and *P. regius*) from non-incubating individuals.

Life at the largest masses

Many factors may limit the maximal size of mammals and birds, including the costs of body maintenance, activity, and reproduction, which all increase with mass. Another factor influencing maximal size is food habits because a sustainable abundance and quality of food is required to support a population of consumers. Still another, of course, is the ability of a consumer to track food resources because as consumer mass increases, the consumer must move greater distances to make up for local limitations in the food supply.

A big appetite

Comparatively large animals have evolved in terrestrial environments, but none have reached the mass of the largest whales. The largest baleen whale is the blue (*Balaenoptera musculus*), which weighs 120–160 t, and the largest carnivorous whale is the sperm (*Physeter catodon*) at 35–50 t, which feeds extensively on squid. The largest marine, vertebrate-eating carnivore is the killer whale or orca (*Orcinus orca*) at 9 t. The largest living terrestrial mammals are the African elephant (*Loxodonta africana*) at 3.0–7.5 t, Indian

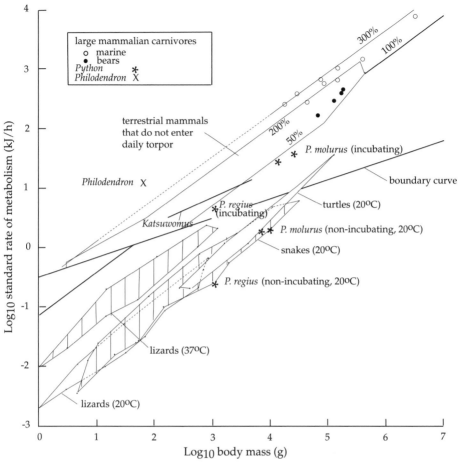

Figure 4.9. Log$_{10}$ basal rate of metabolism as a function of log$_{10}$ body mass in large mammalian carnivores and terrestrial mammals that do not enter torpor. Also indicated are the standard rates of metabolism of reptiles at two temperatures, a tuna (*Katsuwonus*), a thermoregulating plant (*Philodendron*), and a thermoregulating reptile (*Python*). The mammalian basal curves at various levels and the boundary curve are indicated. (Modified from McNab 1983.)

elephant (*Elephas maximus*) at 3.0–5.0 t, white rhinoceros (*Ceratotherium simum*) at 3.5–4.5 t, black rhinoceros (*Diceros bicornis*) at 1.0–1.8 t, and the hippopotamus (*Hippopotamus amphibius*) at 3.0–4.5 t, which is partly aquatic. Some terrestrial mammals of the Pleistocene were about the same size, including mammoths (*Mammuthus* spp.), which weighed 6–8 t, and the mastodon (*Mammut americanum*), at 4–6 t. The largest terrestrial herbivorous mammal known to have existed was a rhinoceros relative, *Paraceratherium* (= *Baluchitherium*) *transouralicum*, which may have weighed 11–15 t (Fortelius & Kappelman 1993).

A species can increase its body size in at least two ways. One is to live in an environment with abundant food resources, and the other is to reduce the cost of body maintenance. A greater abundance and quality of resources in the marine environment appears to account for the larger maximal body masses found in marine vertebrates (McNab 2009c). The abundance and digestibility of plankton are greater than those of terrestrial plants, the most widely used of which is grass, which requires an extended period of fermentation to digest. This difference permitted the larger masses of baleen whales, which in turn permitted the evolution of larger carnivores in the marine environment. The difference in maximal masses between terrestrial and marine environments, when species with a commitment to high energy expenditures (here, mammals) are compared, is approximately 10- to 12-fold.

The other way to increase body mass, as we have seen in the analysis of the boundary curve for endothermy, is to reduce the total cost of body maintenance. A 2.7 g endothermic mammal that conformed to the standard metabolism-mass curve would have the same total rate of metabolism as a 100 mg hypothetical endotherm that conformed to the boundary curve. This pattern is seen when a limit to maximal energy expenditure is dictated by the abundance and quality of resources (see fig. 4.6): that is, if a species or clade conforms to a lower mass-independent energy expenditure, it can reach the maximal energy expenditure dictated by resources at a larger mass than another species or clade that has a higher cost of maintenance. We saw this in the comparison of *Sorex* shrews, crocidurine shrews, and dasyurid marsupials, which form a sequence from high cost of maintenance at small masses to low cost of maintenance at larger masses, all at the same energy expenditure.

This analysis raises the question of whether the largest mammals attained their masses because they have low mass-independent rates of metabolism. Unfortunately, few standard and no field measurements of metabolism are available for the largest mammals. The largest mammal to yield a reasonable estimate of its basal rate is probably the killer whale; (Williams et al. 2001); that estimate is 227% of the value expected from mass at 3.22 t. Some of the other large mammals measured are the moose (*Alces americana*), with a BMR at 209% of the expected value at 350.0 kg; Weddell seal (*Leptonychotes weddelli*), with 200% at 388.5 kg; and dromedary camel (*Camelus dromedarius*), with 104% at 407.0 kg; the lower rate in the dromedary may reflect its desert existence (see chap. 7). Benedict (1938) measured the rate of metabolism of Indian elephants (*Elephas maximus*), which is 225% at 3.67 t. Data on large herbivores may or may not be good estimates of basal rates, given the concern over whether gut fermentation ever permits these species to be postabsorptive. The few available data appear to indicate that the largest mammals did not attain their masses through a reduction in mass-independent energy expenditure. Furthermore, large marine carnivores have especially high BMRs (see fig. 4.9), which may be responses to a cold aquatic habitat (see chap. 3) as well as to a high resource base and a high activity level.

The largest mammals in terrestrial environments were exceeded in mass by the largest herbivorous and carnivorous dinosaurs. The largest herbivorous dinosaurs weighed between 40 and 70 t, although *Argentosaurus* may have weighed up to 100 t (Appenzeller 1994). The largest carnivorous dinosaurs included *Tyrannosaurus* (4–8 t), *Szechuanosaurus* (7–10 t), and *Deinochirus* (7–10 t) (Peczkis 1994). If large dinosaurs were inertial homeotherms (McNab 1978a; Gillooly et al. 2007), their masses may have reflected a low cost of body maintenance. For example, if one assumes that the field energy expenditure of a 7.5 t African elephant expected from extrapolating the field energy expenditures of mammals (Nagy et al. 1999; see chap. 11) represents the limit on energy expenditure for herbivores set by resources in a terrestrial environment, that limit would be about 5.4×10^5 kJ/d (fig. 4.10). If, however, herbivorous dinosaurs encountered the same limit

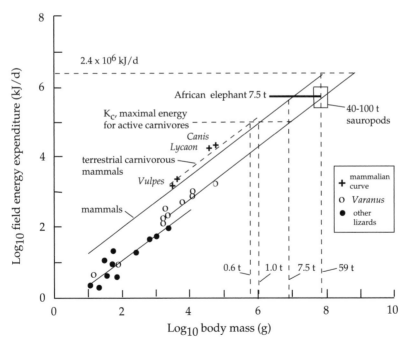

Figure 4.10. Log_{10} field energy expenditures as a function of log_{10} body mass in mammalian carnivores, the African elephant (*Loxodonta africana*), and various lizards, as well as an estimate for 40–100 t sauropods. (Modified from McNab 2009c.)

and had field energy expenditures, corrected for mass, that are similar to those of varanid lizards, which are about one-fifth those of mammals, those expenditures and that limit would permit a body mass of about 59 t (McNab 2009c). On the other hand, if the energy expenditure of a 59 t dinosaur conformed to the mammalian curve, it would have expended 2.4×10^6 kJ/d, or 4.4 times the field energy expenditure expected in the elephant. Thus, herbivorous dinosaurs may have attained maximal masses some eight times those of mammals as a result of having field energy expenditures that were about 22% of those of mammals (McNab 2009b). (Note that $[8^{0.72}]^{-1} = 0.22$; in other words, these two estimates, of size and metabolism, are consistent [see chap. 11].)

As was seen in the comparison of terrestrial and marine mammals, the large size of dinosaurs might also have reflected a greater resource base in the Mesozoic than presently seen in East Africa. However, the Mesozoic terrestrial flora principally consisted of ferns, conifers, cycads, and *Ginkgo*; grasses, an exceedingly important resource for large mammalian herbivores today, were absent (Weaver 1983). Therefore, it appears unlikely that the masses of large herbivorous dinosaurs reflected a greater terrestrial resource base than is found today (Hummel et al. 2008); if any difference existed, the resource base in the Mesozoic may have been poorer.

This analysis implies that terrestrial plants are limited in their ability to sustain the energy expenditures of consumers, which may require large consumers to limit their expenditures in turn by limiting their body size or reducing their mass-independent expenditures (McNab 1978a, 2009c). A parallel situation exists on islands, where large mammals, because of their commitment to endothermy and its high expenditures, tend to become smaller when exposed to an area-based restriction on energy availability (see chap. 9). However, Köhler (2010) showed that a bovid (*Myotragus* spp.) apparently responded to life on Majorca with a reduction in both its body mass (ca. 19 kg) and its mass-independent rate of metabolism to the extent that its commitment to endothermy was, at least, relaxed (as shown by annual growth rings in its skeleton).

Birds are obviously limited to smaller masses than mammals, principally because of their commitment to flight. Among living birds, the largest flying species are the wandering albatross (*Diomedea exulans*, 6–11 kg), Andean condor (*Vultur gryphus*, 8–15 kg), California condor (*Gymnogyps californicus*, 8–14 kg), Kori bustard (*Ardeotis kori*, 6–19 kg), Eurasian bustard (*Otis tarda*, 4–18 kg), and mute swan (*Cygnus olor*, 7–15 kg). The largest flying bird may have been a teratorn, *Argentavis magnificens* (ca. 70 kg?) from the Miocene of Argentina (Campbell & Tonni 1983; Palmqvist & Vizcaíno 2003).

When the commitment to flight is abandoned, avian body mass may increase appreciably. Thus, the largest living birds are the ostrich (90 –156 kg), emu (*Dromaius novaehollandiae*, 30 –55 kg), southern cassowary (*Casuarius casuarius*, 29–58 kg), and greater rhea (*Rhea americana*, 20 –25 kg). The largest parrot is New Zealand's flightless kakapo (*Strigops habroptilus*, 1–3 kg), and the largest rail is the flightless takahe (*Porphyrio mantelli*), also from New Zealand, which weighs 2 –3 kg. Among fossil species, the largest flightless species included a moa (*Dinornis giganteus*, 100 –200 kg) in New Zealand, the elephant bird (*Aepyornis maximus*, ca. 400 kg) in Madagascar, and the thunderbird (*Dromornis stirtoni*, 500 kg?) in Australia. The famous flightless dodo (*Raphus cucullatus*) of Mauritius was appreciably smaller (ca. 23 kg), but considering that Mauritius is a small island (1,860 km^2), the dodo was quite large. The evolution of a flightless condition in birds decreases mass-independent BMR to near the values expected from mass in mammals (McNab 2009a) (see chap. 3).

Heat matters

A physiological problem for large endotherms in a warm terrestrial environment is overheating. All of the largest living terrestrial mammals are naked, including elephants, rhinos, and the hippopotamus, most of which live in warm climates. The naked condition in hippos, however, is complicated by their aquatic habits; hippos are terrestrial principally at night, when they avoid an intense solar input. During the Pleistocene, several of the largest terrestrial mammals in the Northern Hemisphere, including mammoths, mastodons, woolly rhinos, and ground sloths, had well-developed fur coats. Obviously, in these cases, the fur coats were responses to cold ambient temperatures, and their absence in tropical equivalents is a means of diminishing heat storage. It is possible, however, at least in the case of ground sloths, that a well-developed fur coat may have been compensatory for a low level of energy expenditure, as occurs in their intermediate-sized arboreal relatives, *Bradypus* and *Choloepus* (see fig. 4.2) (McNab 1978c), and in the binturong (*Arctictis binturong*) (McNab 1995).

High ambient temperatures may limit body size. The large masses of some marine mammals may partially reflect the absence of a heat load, but also depend on an abundance of resources, especially in polar environments. In a hot environment, the necessity to dump the heat produced by metabolism may limit body size, especially in species with high metabolic rates; this observation may be further evidence of low metabolic rates in dinosaurs. An example of high-temperature reduction of body size and rate of energy expenditure occurs at the opposite end of the size scale, in the fascinating case of the naked mole-rat (*Heterocephalus glaber*) of East Africa. As we shall see (in chap. 7), it lives in closed burrows that often have temperatures that vary from 29°C to 34°C with an atmospheric relative humidity between 92% and 100% (McNab 1966). The mole-rat has responded to these oppressive conditions by decreasing its body mass to 30–60 g, becoming the smallest bathyergid, and it has so increased its thermal conductance and reduced its basal rate that it has nearly abandoned endothermy.

CHAPTER FIVE. *A Diversity of Food Habits*

One of the pivotal biological relationships is that between species and their food supplies, from which animals obtain their nutrients and the chemical potential energy they need for growth, development, maintenance, and reproduction. Therefore, the "choice" of foods is likely to be a pivotal component of any species' natural history and specifically of its energy expenditure. As we saw in chapter 3, BMR in both mammals and birds correlates with food habits. A more subtle impact of food habits on BMR may be seen in groups of related species that have diversified food habits; this comparison minimizes the potential effects of other factors that might also influence energy expenditure, including evolutionary history. An examination of the basal rates of unrelated species that have converged on a food would also help to establish the degree to which food habits influence BMR, or whether a response is limited by affiliation with a particular evolutionary line.

The impact of food habits in related species

Mammals

The first attempt to determine whether the residual variation in the BMR of mammals correlated with food habits involved the Neotropical bat family Phyllostomidae (see box 5.1). These bats have the greatest diversity in food habits of any vertebrate family, including species that feed on insects, fruit, nectar/pollen, and vertebrates (including mammals, birds, and frogs), both as pure diets and in various combinations. The three species of vampire bats are now judged to be members of this family (Wetterer et al. 2000), which adds avian and mammalian blood to its dietary repertoire. Another family of interest was the Noctilionidae, which has two species, a smaller insectivore (*Noctilio albiventris*) and a larger insectivore/piscivore (*N. leporinus*).

Measurements on the phyllostomids, vampires, noctilionids, and a few insectivorous molossid and vespertilionid bats clearly indicated that much

BOX 5.1. ....

Studying Bats in Brazil

As a finishing graduate student, I was invited by Peter Morrison to participate in an expedition to study the adaptation of mammals to high altitudes in Chile and Peru in 1959–1960. Peter had decided to start by spending a month at Hospital das Clinicas in Bahia (Salvador), Brazil, and then another month at the Universidade de São Paulo in the physiology department of Professor Paulo Sawaya. I was advance person on the expedition and set up working conditions in Bahia. While waiting a month for the other members of the expedition, I mist-netted bats for study. As a result, I caught many fruit bats belonging to the family Phyllostomidae. As it turned out, we had difficulty getting some of our equipment through Brazilian customs, so we lost the opportunity to measure the metabolism of these bats.

In 1966–1967, I returned to Brazil for 15 months to measure the energy expenditure of bats while resident in the Departamento de Fisiologia Geral e Animal at the Universidade de São Paulo. I was especially interested in the Phyllostomidae, including the three species of vampire bats. Most of my work was done in central São Paulo State, but work on *Noctilio* was accomplished in the interior, near Rio Claro, and some collecting of bats was done in the adjacent state of Minas Gerais.

When I was in São Paulo, Pete Scholander came though the department to coordinate a visit to the Amazon basin by the *Alpha Helix*, a research ship that was part of the fleet at Scripps Institution of Oceanography. Sawaya was the Brazilian coordinator of the expedition. Scholander, whom I had met at Madison, Wisconsin, when I was a graduate student, asked if I would like to work on the ship when it went up the Rio Negro. I jumped at the chance because it was going to permit me to work on a greater array of phyllostomids.

I received tickets from Scripps to fly to Manaus to meet the *Alpha Helix*. I checked into a hotel and went to the dock, found the *Helix*, and asked to see Knut Schmidt-Nielsen, who was the senior scientist on one of four excursions, this one up the Rio Negro. I met Knut, who was not very hospitable, to say the least! In a day or so, I flew up the Rio Negro on a floatplane to stay at a shore camp. Knut and several others also stayed there for part of the time because they wanted to have wine for dinner, but the captain of the ship refused to allow it, except for Sunday afternoon, when everyone was under orders to congregate on the poop deck with beer, wine, and goodies. I was the only person on this excursion who regularly went into the surrounding forest. (I had to go upstream by outboard in case I broke a cotter pin because I could float down to the *Helix*—if I had gone downstream and broken a pin, it would have been a long float down to Manaus.) I was mist-netting bats in the forest at night and was able to capture two *Diaemus youngi*, a rare vampire that I had given up hope of catching. I invited other members of the expedition to come with me to check the nets, and most, including Knut, did, but there was one biochemist who never left the ship, except for a 10-minute ride to the shore camp just before the ship returned to Manaus: the "jungle" was dangerous!

At the end of the trip, Knut kindly thanked me for coming on the trip. Later I saw him at Duke and he was very kind, reminiscing about our time on the Rio Negro, and

I was included in his memoir (1998). What I later found out was that Scripps had invited Sawaya, as Brazilian coordinator, to name three Brazilians to be part of the expedition, and he had named me as one. This information was communicated to Scripps, who sent the information to Knut. Knut made the point that he was chief scientist on this expedition and that I was not a Brazilian and could not be part of the excursion. One simply had to know Sawaya! He pointed out to Scripps that he was the Brazilian coordinator of the *Alpha Helix* trip, and that if he did not approve, the ship could not enter Brazilian waters! I was part of his Brazilian contingent, period. I received tickets to Manaus from Scripps.

. . . .

of the residual variation in the basal rates of tropical bats, after accounting for the impact of body mass, correlated with food habits (McNab 1969). Data on 22 species of phyllostomids indicated that the 14 that fed on vertebrates, nectar, or fruit had high basal rates; 3 species of *Phyllostomus* with mixed diets had intermediate basal rates; and 5 species that fed on flying insects or blood had low basal rates. So, the two factors that appeared to be most influential in setting the basal rate in these tropical bats were body mass and food habits (McNab 1969, 1989), although this conclusion was made at a time when I was not able to make a more detailed analysis.

Cruz-Neto and colleagues (2001) challenged this analysis thirty-two years later: they suggested that the various levels of energy expenditure in this family were not correlated with food habits, but rather reflected "recency of ancestry" (see chap. 2). In response, I performed a more elaborate analysis of phyllostomids (McNab 2003b), this time with 30 species (fig. 5.1A). The new analysis indicated that body mass accounted for only 78.7% of the variation in BMR, in part because the range in mass was only by a factor of 11:1 and in part because of the large residual variation. (Remember that the great range in mass is why mass is so effective in accounting for the variation in BMR in the class Mammalia, as we saw in chapter 1; obviously, if a character varies among several species of identical size, none of the variation is due to mass.)

The basal rates of phyllostomids correlated with three factors other than mass: food habits, elevation, and distribution. As we have seen, one difficulty is to decide on the number of categories into which to divide food habits. If ten food categories were used, the four factors accounted for 99.4% of the variation in basal rate. An interesting result of this analysis was the impact of eating different kinds of fruits: four species that specialized on *Piper/Vismia/Solanum* fruits had a dimensionless food coefficient equal to 1.00, eight species that specialized on *Ficus* had a coefficient equal to 0.84, and one species that specialized on Guttiferae fruits had a coefficient equal to 0.75. All fruits

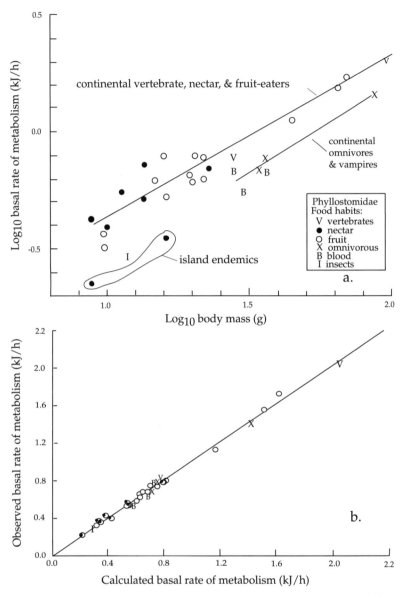

Figure 5.1. (A) Log$_{10}$ basal rate of metabolism as a function of log$_{10}$ body mass in phyllostomid bats in relation to their food habits and distribution. (B) Measured basal rates of metabolism of phyllostomid bats as a function of basal rate of metabolism calculated from equation (5.1). (Modified from McNab 2003b.)

obviously are not equal. However, not all of the ten food categories had statistically different mean basal rates, so those whose basal rates were not different were combined, which resulted in four food categories: nectar, fruit and vertebrates, omnivory, and blood and insects. The coefficients for these food habits equaled 1.49 for nectarivores, 1.35 for frugivores and carnivores, 1.00 for omnivores, and 0.92 for insectivores and vampires (note that the latter two categories were only marginally different). With this analysis, continental species had a mean BMR that was 68% greater than that of island endemics ($P < 0.0001$), and highland species had a mean basal rate 39% greater than that of lowland species ($P = 0.0013$). Therefore,

$$\text{BMR (kJ/h)} = 0.080\,(D \cdot E \cdot F)\,m^{0.737}, \tag{5.1}$$

where D, E, and F are dimensionless coefficients that stand for distribution, elevation, and food habits, respectively, and $r^2 = 0.972$. The degree to which food habits were subdivided obviously influenced r^2. Clearly, the vast majority of the quantitative variation in BMR in phyllostomids is accounted for by this analysis, which is an improvement over a Kleiberan analysis when $r^2 = 0.787$. (fig. 5.1B).

Whether food habits influenced the BMR of other mammals, or of other endotherms in general, was unexamined the time of the phyllostomid study. That study encouraged the examination of other trophically diverse mammals (for a more general view, see chap. 3). In a sense, it was part of a minor intellectual breakthrough that actually began with a demonstration that much of the residual variation in *Peromyscus* BMR correlated with climate (McNab & Morrison 1963). Most biologists in the 1960s ignored the residual variation around the metabolism-mass curve, presumably assuming that it was measurement error and could therefore be reduced by improved measurement techniques. However, the important clue was that the data on *Peromyscus* and on phyllostomids indicated that the species falling above the scaling curve were ecologically or trophically more similar to one another than they were to those that fell below the curve, and a similar statement held for the species that fell below the curve. Therefore, some factors other than mass were determining the residual variation in BMR, at least in these examples.

The two noctilionids had slightly different basal rates: the smaller insectivorous species (*N. albiventris*, 27 g) had a BMR equal to 64% of the value expected from mass, whereas the larger *N. leporinus* (61 g), which feeds extensively on fish and insects (Novick & Dale 1971), had a slightly higher BMR, at 69%. This observation raises the question, given the BMRs of carnivorous phyllostomids, why *N. leporinus* had such a low BMR. In contrast, the phyllostomid *Chrotopterus auritus* (96 g), a hard-core carnivore, had a basal rate equal to 109% of the value expected from mass, and *Tonatia*

BOX 5.2. ⋯⋯⋯⋯⋯⋯⋯⋯⋯⋯⋯⋯⋯⋯⋯⋯⋯⋯⋯⋯⋯⋯⋯⋯⋯⋯⋯⋯

Food Habits of Phyllostomid Bats

Maintaining bats in captivity taught me a lot about their food habits. I once maintained the committed vertebrate eater, *Chrotopterus auritus* (96 g), which ate white mice in captivity. I also captured *Tonatia bidens* (27 g), which was reputed to be insectivorous. Realizing that insectivorous bats are difficulty to keep in captivity, I gave *Tonatia* a white mouse out of desperation. It immediately recognized the mouse as prey, jumping on its back, biting it on the back of the skull, and consuming everything but the tail. Recent information in the field has shown that this species is a bird eater (Martuscelli 1995). Once I accidentally placed a vampire (*Desmodus rotundus*) in a cage with a *Chrotopterus*, which immediately attacked, killed, and ate the vampire. This aggressive behavior explains why *Chrotopterus* roosts alone.

In Bahia, I had several cages of bats, one of which contained three or so of the large (84 g) *Phyllostomus hastatus*, which were caught at fruiting trees and were maintained in captivity on fruit, and another that contained several small, insectivorous (16 g) *Molossus molossus*. The *Molossus* cage was placed above the *Phyllostomus* cage. One afternoon I finished some measurements on a *Molossus* and, in a rush, accidently put it into the *Phyllostomus* cage. I turned to remove another *Molossus* from an experimental chamber and heard a crunching, only to turn around to see that a *Phyllostomus* had the *Molossus* in its mouth and was chewing on it. When the *Phyllostomus* was finished, only the wing tips were left. *Phyllostomus hastatus* is obviously an omnivore!

Diphylla in captivity would not consume cattle blood obtained from a slaughterhouse, a food readily eaten by *Desmodus*. When I placed a live chicken in a cage with *Diphylla*, it lowered itself onto the chicken's back and swam through the feathers to feed around the cloaca, or it would walk on the floor, like a large spider, to nibble at the chicken's toes. It clearly was a bird vampire.

⋯⋯⋯⋯⋯⋯⋯⋯⋯⋯⋯⋯⋯⋯⋯⋯⋯⋯⋯⋯⋯⋯⋯⋯⋯⋯⋯⋯⋯⋯⋯⋯

bidens (27 g), with a mixed insect/vertebrate diet, had a basal rate equal to 104% (see box 5.2). Why does *N. leporinus* have a lower mass-independent basal rate? A greater dependence on insects? (Measurements on the frog-eating *Trachops cirrhosus* [32 g] and the largest carnivorous phyllostomid, *Vampyrum spectrum* [150 g], as well as on other species of the genus *Tonatia*, would be very valuable.)

Members of the order Carnivora also have a diversity of food habits. Many of these animals do not eat vertebrates (i.e., are not carnivorous), including some that eat insects and others that consume fruit and leaves. Frugivorous/folivorous carnivorans are found mainly in the tropics; they include viverrids from Southeast Asia, nandinids from Africa, and procyonids from the Neotropics as well as the bamboo-eating panda (*Ailuropoda melanoleuca*) and the fruit- and leaf-eating red panda (*Ailurus fulgens*) from

BOX 5.3. ··

Non-carnivory in the Binturong

My attempt to feed a 15 kg frugivorous binturong (*Arctictis binturong*), a member of the Viverridae and order Carnivora, a 15 g white mouse was marginally successful: it was frightened by the live mouse, but after I killed the mouse, the binturong tried to pull it apart, as if it were a piece of taffy, while stepping on the carcass. It vaguely knew that the mouse was edible, but was completely incapable of being an effective predator on a mouse 1/1,000 of its size!

··

Asia (see box 5.3). As we saw in chapter 3 (see fig. 3.2), these species have low basal rates that are associated with their food habits and arboreal lifestyles.

The first measurements on vertebrate-eating carnivorans were problematic: they were derived mainly from mustelids (i.e., weasels, stoats, and ermine), but included a few questionable high measurements on seals. The difficulty with mustelids was that they have a long body, which potentially meant that their high BMRs might reflect high rates of heat loss associated with a large surface area, rather than their food habits. Furthermore, the seals may not have had their actual basal rates measured, and besides, they live in cold water, which could have been a climatic basis for a high basal rate.

Whether a high basal rate is associated with vertebrate eating per se might be determined by measuring the basal rates of committed terrestrial carnivores that have a more normal mammalian shape. That leaves out canids, most of which are trophic opportunists that feed on a variety of foods, with the exception of the Holarctic wolf (*Canis lupus*), which has a high basal rate (128% [Okarma & Koteja 1987]); the African hunting dog (*Lycaon pictus*), which has a high rate of metabolism (242%), but was measured under nonstandard conditions (Taylor et al. 1971); and the Southeast Asian dhole (*Cuon alpinus*), which has not been measured. What options were left? Cats! Most felids are strict carnivores, and they live in a variety of environments, but the only felid that had been measured by 1990 was the house cat (*Felis sylvestris* [*domesticus*]), which also has a high basal rate (125% [Benedict 1938]). The hint from these data was that vertebrate-eating eutherian specialists indeed have high basal rates, but more data were needed to be sure of this conclusion.

I was able to borrow nine species of felids (and a striped hyena [*Hyaena hyaena*]) to measure their rates of metabolism as a function of ambient temperature (McNab 2000c) (see box 5.4). These felids included two bobcats (*Lynx rufus*), two ocelots (*Leopardus pardalis*), two margays (*Leopardus wiedii*), two jaguarundis (*Puma yagouaroundi*), four pumas (*Puma concolor*),

BOX 5.4. ·

Working with Large Cats

I was able to work with large cats as a result of the cooperation of many people and institutions, including the teaching zoo at Santa Fe College, Gainesville, Florida; Parque Ambiental y Zoológico Gustavo Rivera, Falcón, Venezuela; Central Florida Zoo, Lake Monroe, Florida; the Veterinary School, University of Florida; and the U.S. Fish and Wildlife Service. One day, out of curiosity, I drove to what a highway sign indicated to be the "Wild Animal Retirement Village," which was just a few miles away from Gainesville, Florida. When I got there, I found a variety of animals housed by a couple, Gene and Rusti Schuler, who were retired circus members. They had decided to dedicate the rest of their lives to taking care of wild animals that were no longer wanted. Some people, for example, fell in love with a lion kitten, which later grew to be a 150 kg adult that possibly wanted the living room couch for itself and the right to chose the television channel! Or a captive leopard might be constantly trying to capture its owner. The animals were of many ages, including puma kittens that were born in the village. Some of the adult pumas would purr when petted. As soon as I saw this opportunity, I asked whether I might be able to borrow some cats to measure their rates of metabolism.

Gene said that yes, we could try. He brought "Sandy," a 98 kg female lion, over to the university. Gene parked behind Bartram Hall, where the zoology department was housed, and we had to take the lion up to the sixth floor, where my laboratory was located. He said that the animal was very nice and that we could take her on a leash. Well, we had two choices: take her into the elevator or walk her up the stairwell. I thought it would be best to walk up because that way we three would not be enclosed in a small elevator. (See, I am cautious!) So, we walked Sandy up seven floors, but on the way we met a class that had just finished and was walking down the stairs between the first and second floors. The students scattered, while Gene shouted: "Careful, this is a dangerous animal!" Oh?

I kept the lion in Jack Kaufmann's office because he, a behavioral ecologist in the department, had some large cages in which he had kept some procyonids. The next day, I went down to Jack's office to walk Sandy to my lab by leash—my first error. I walked her down the hallway, watched by a group of graduate students standing near the elevator, who were buzzing about the lion. I got to my laboratory and tried to get Sandy to enter a large cage on wheels, which I planned to roll into a temperature-controlled chamber to measure her rate of metabolism. But Sandy would not go into the cage, even though I pushed her on the rear. Thus, I made my second error: I got into the cage and tried to pull her in. Sandy came in halfway, I unhooked the leash, and she turned around, left the cage, and because I had not closed the door to my lab—maybe that was my second error—trotted down the hallway toward the graduate students. So, I ran after Sandy, jumped on her back, pulled her down on top of me, and asked my Swiss postdoc, Michel Genoud: "Get the leash!"

I had an opportunity to measure an adult male tiger, which weighed 140 kg. Gene told me that "Orange" was a very nice animal, but that I should remember that he was "fully equipped," meaning that he had his teeth, claws, and testes. He was a sweet animal. But one day I went down to get him from Jack's office—now animals were placed in the wheeled cage and rolled to my office. (I am capable of learning.) As Orange was being guided to the cage, he heard a noise in the hallway. Dr. Paul Chun, a Korean biochemist, who had the laboratory next to Jack, was writing some formulae on a chalkboard in the hall. Orange was curious to know what was making the noise in the hallway, so he jumped onto a sink that was just inside Jack's doorway and stuck his head out to see what was going on. I thought that Dr. Chun was just about bite size for Orange, so I grabbed his tail, pulled him off the sink, and pushed him into the cage. Nevertheless, Orange was a very nice animal, and he permitted me, as did most of the cats, to measure his body temperature by inserting a thermocouple into his rectum after an experiment.

But not the jaguars! They were treated with respect: whereas all of the other species had their rectal temperatures measured after an experiment, the jaguars were never touched. The first one I worked with was a male, which never looked at me, but as soon as my back was turned, went after me: by design, he was in the cage and I was outside. On the first night this jaguar was wheeled into the metabolism chamber, I looked into the chamber just before I left and saw that he was mouthing the lock that secured a chain and the guillotine door of the cage. When I returned in the morning and peeked in, the jaguar, unbelievably, had opened the cage and was sitting on its top. (I must have not completely closed the lock out of concern that the jaguar might rush me; I am sure that the jaguar did not know the lock's combination.) I called Gene, and he said "Don't do anything!" He came over and tried to push the animal off the cage by slipping a two-by-four into the chamber through the partly opened door, but instead the jaguar tried to walk out on it. So, we pulled out the two-by-four, closed the chamber, and called the veterinary school, and a vet came over and tranquilized the cat. Later I got the animal back and got some good measurements on him, and on an equally devious female. Being a comparative biologist is an adventure.

. . . .

a cheetah (*Acinonyx jubatus*), two jaguars (*Panthera onca*), a lion (*Panthera leo*), and a tiger (*Panthera tigris*). The results showed that all species but two had high basal rates (114%–165% of the values expected from mass), which suggested that the high BMRs of weasels really reflected their food habits (fig. 5.2). However, the jaguarundi and margay had low mass-independent BMRs, 74% and 73%, respectively. This observation is especially interesting because both species belong to genera containing other species that had high BMRs. Such is the problem and thrill of this work: a problem because one must be sure that these measurements are reliable, and a thrill because these deviant measurements tell us something about the factors that contribute to biological diversity. But what?

The low basal rate of the margay is especially noteworthy. This cat is very slim, without the bulk found in its congener, the ocelot, a muscular species. The major ecological difference between these two species is that the

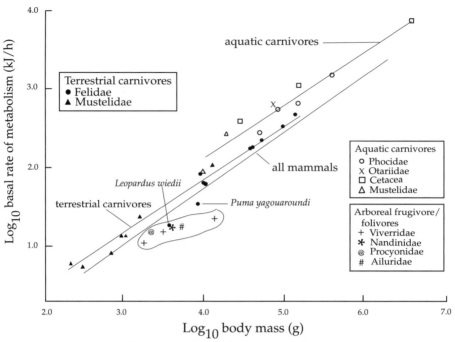

Figure 5.2. Log_{10} basal rate of metabolism as a function of log_{10} body mass in members of the order Carnivora in relation to an aquatic, terrestrial, or arboreal frugivore/foliovore existence. Data on Cetacea are included. (Data from McNab 2005c.)

margay is principally arboreal, whereas the ocelot is principally terrestrial. Arboreal mammals are generally characterized by low basal rates, which in part may be associated with a lean body composition, so the margay's slim body may have accounted for its low basal rate. This explanation could be tested, either by an anatomical dissection (does it have an unusually small muscle or heart mass for a felid?) or by examining the energy expenditure of another tropical arboreal cat, the marbled cat (*Pardofelis marmorata*) from Southeast Asia, a species that has not been measured. Either test, however, would leave the low basal rate of the jaguarundi unexplained. The jaguarundi is a long, slender, terrestrial species whose biology is little known, but it is an agile climber. It is apparently committed to feeding on vertebrates, including small mammals, birds, and reptiles (Tófoli et al. 2009).

Another group of carnivores worthy of consideration is the pinnipeds, which were measured by Irving et al. (1935), Scholander (1940), Irving and Hart (1957), Hart and Irving (1959), Kooyman et al. (1973), Iversen and Krog (1973), Miller and Irving (1975), and Miller et al. (1976). All of these studies indicated that pinnipeds have much higher rates of metabolism than expected from mass. Doubt has been raised, however, as to whether these measurements represented standard conditions (Lavigne et al. 1982; Schmitz & Lavigne 1984; Lavigne et al. 1986; Boyd 2002), or whether the high rates reflected anxiety, forced restraint, or immature individuals. Some reports indicated that seals had rates that conformed to the Kleiber curve (Lavigne et al. 1986), although later measurements found high basal rates (Boily & Lavigne 1997; Costa & Williams 1999; Costa 2001; Williams et al. 2001; Boyd 2002; Williams & Worthy 2002).

The latest measurements on three cold-water pinnipeds, *Pagophilus groenlandicus, Phoca vitulina*, and *Pusa hispida*, demonstrated that they do indeed have high basal rates (180%, 207%, and 181%, respectively [Ochoa-Acuña et al. 2009]). As noted, the high basal rates of terrestrial carnivores—felids and mustelids—are lower than those found in cold-water seals, which in turn have basal rates lower than those found in delphinids (*Tursiops truncatus* [310%], *Orcinus orca* [227%]) and phocoenids (*Phocoena phocoena* [341%]) (see fig. 5.2). The level of energy expenditure in this collective is potentially explained as a response both to carnivory and to life in cold water, as well as by the high activity levels of cetaceans. Knowing whether tropical monk seals (*Monachus monachus* and *M. schauinslandi*) or the Antarctic crabeater seal (*Lobodon carcinophagus*), which feeds principally on krill, have equally high basal rates would shed light on the extent to which high basal rates reflect life in cold water, activity, or the use of fish as prey. Again, we are faced with the superficiality of correlations, but also with the realization that by pursing additional leads, these relationships can be clarified.

An analysis of the factors that determine the basal rates of 62 carnivorans showed that body mass alone accounted for 86.8% of the variation in BMR (McNab 2005c), which left room for other factors to influence carnivoran BMR. Other factors with which BMR correlated included substrate (S), food habits (F), habitat (H), and latitude (L):

$$BMR \text{ (kJ/h)} = 3.16 \ (S \cdot F \cdot H \cdot L) \ m^{0.694}. \tag{5.2}$$

This relationship accounts for 98.7% of the variation in carnivore BMR (table 5.1) and even for the variation in the basal rates of cetaceans, which were not included in the analysis (fig. 5.3). With regard to food habits, folivorous, insectivorous, and frugivorous species collectively had basal rates that averaged 70% of the mean for species that were carnivores or omnivores (see fig. 5.2). Therefore, according to the analysis represented by equation (5.2), arboreal, forest-dwelling, tropical frugivores would be expected to have basal rates that were $0.82 \times 0.70 \times 0.79 \times 0.82 = 0.372$; that is, only 37.2% of the basal rates of terrestrial, grassland, temperate carnivores. A direct comparison of these two groups indicates that the frugivores actually had a mean basal rate that was 46.7% ($n = 7$) of the mean of the carnivores;

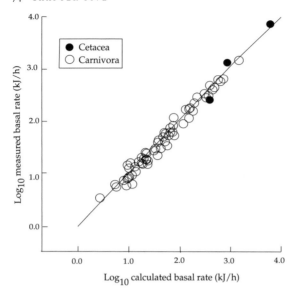

Figure 5.3. Measured basal rates of metabolism in Carnivora and Cetacea as a function of the basal rates calculated from equation (5.2).

TABLE 5.1. Dimensionless coefficients for the ecological characteristics of carnivorans.

Factors	Categories	Value
S: Substrate	Burrows/arboreal	0.82
	Terrestrial	1.00
	Aquatic	1.82
F: Food	Leaves, insects, fruit	0.70
	Other foods	1.00
H: Habitat	Forests, deserts	0.79
	Other habitats	1.00
L: Latitude	Temperate	1.00
	Other latitudes	0.82

in other words, the estimated depression of BMR in these arboreal species was exaggerated. Some of this error may reflect the combinations of some character states, such as burrowing and arboreal lifestyles (see table 5.1).

The giant panda (*Ailuropoda melanoleuca*), a semiarboreal, bamboo-eating bear, has not been measured, but given its habits, it undoubtedly has a BMR that is appreciably less than expected from mass. Its basal rate is possibly similar to, or less than, that found in the sloth bear (*Melursus ursinus*), which feeds heavily on termites and has a basal rate equal to 80% (McNab 1992a). The sloth bear, and presumably the giant panda, differ in energetics

from the omnivorous black (*U. americanus*, 90% [Watts & Cuyler 1988])
and brown (*U. arctos*, 96% [Watts & Jonkel 1988]) bears and the strictly
carnivorous polar bear (*U. maritimus*, 103% [Watts et al. 1987]).

An extensive examination of energetics in Neotropical caviomorph
rodents was made to determine the extent to which variation in BMR
depended on body size, food habits, substrate, and distribution (Arends &
McNab 2001). Among 30 species, BMR varied principally with body mass
($r^2 = 0.945$), as expected given this group's large range in body size (100 g to
80 kg). BMR in caviomorphs also correlated with substrate and mass ($r^2 =
0.969$), but not with food habits as long as substrate, climate, distribution on
islands or continents, or familial affiliation was included. The caviomorphs,
as diverse as they are in the number of species (ca. 200) and families (10),
are narrowly herbivorous. So, it is not surprising that BMR is a significant
correlate of food habits in caviomorphs only when combined with body
mass alone. Then, folivorous species have basal rates that are 71% of those of
species with other herbivorous habits (fig. 5.4). The reason why food habits
cannot be combined with other factors is factor interaction: three of the
five folivores are island endemics; most of the grazers live on continents; all
of the folivores are arboreal; and sixteen of the eighteen grazers are terrestrial, the other two being aquatic. These interactions make the assignment

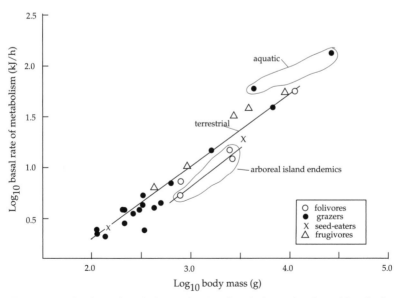

Figure 5.4. Log_{10} basal rate of metabolism as a function of log_{10} body mass in rodents of the suborder
Caviomorpha in relation to food habits, a terrestrial or aquatic habit, and a continental or island
distribution. (Modified from Arends & McNab 2001.)

of responsibility for the determination of caviomorph basal rates difficult to determine.

In this context, the determination of BMR in marsupials is especially interesting. As we have seen, marsupial grazers and carnivores have at best intermediate mass-independent basal rates (McNab 2005b), whereas eutherians with these food habits have high mass-independent basal rates (McNab 2008a). Therefore, the range in mass-independent basal rates in marsupials with reference to food habits is much narrower than that found in eutherians (see fig. 4.5) (see also chap. 12). As a consequence, marsupial BMRs are not correlated with food habits ($P = 0.27$) when coupled with body mass (McNab 2005b).

Birds

Food habits also appear to influence the basal rates of birds, although this topic has been less explored than in mammals. A general analysis of avian energetics (McNab 2009a) indicated that species that feed on nuts, nectar, or a mixed diet with insects have the highest basal rates (see chap. 3), followed by species that feed on insects, vertebrates, aquatic invertebrates, or seeds, and finally by those that feed on fruit (McNab 1988b, 2009a). The extent to which food habits influence BMR in birds is quantified in table 3.2.

Two examples of the influence of food habits on BMR within a family of birds are the rails (family Rallidae) of Oceania (McNab & Ellis 2006) and the birds of paradise (Paradisaeidae) of New Guinea (McNab 2003c, 2005a). Three factors were important among eleven species of rails: body mass, food habits, and a flighted or flightless condition (McNab & Ellis 2006) (see chap. 9). Species that are omnivorous had basal rates that were 73% of those of species that feed on vegetation (i.e., gallinules). With regard to birds of paradise, BMR was measured in thirteen species, although two species, Loria's "satinbird" (*Cnemophilus loriae*) and the recently studied crested "satinbird" (*C. macgregorii*), are no longer considered to belong to this family, having been placed in a separate family, the Cnemophilidae (Cracraft & Feinstein 2000; Irestedt & Olson 2008). Among birds of paradise, BMR correlated with food habits and altitudinal distribution (see chap. 6). Three food habits statistically collapsed into two when food habits were combined with body mass alone. When altitude was brought into the analysis, however, all three were significantly different. This means that food habits and altitude were not independent of each other: committed frugivores in this study were limited to intermediate and high altitudes. Frugivores then had basal rates that were 76% of those of omnivores, and insectivores had basal rates that were 91% of those of omnivores (see chap. 6).

The impact of food habits in convergent taxa

Another approach to examining the impact of food habits on metabolism is to compare the mass-independent energetics of unrelated species that have converged on the same food habits. Such a comparison may permit us to determine the extent to which similar food habits require a similarity in energy expenditure and the extent to which ancestry (phylogeny) may limit this convergence.

A strange mixture of retrograde mammals evolved a distinctive food habit, ant and termite eating (McNab 1984, 2000a) (see box 5.5). Species with this specialized food habit are found throughout the tropics, including tamanduas (Myrmecophagidae) and some armadillos (Dasypodidae) in the Neotropics; pangolins (Manidae) in Asia and Africa; elephant shrews (Macroscelitidae), the aardwolf (*Proteles cristatus*), and the aardvark (*Orycteropus afer*) in Africa; tenrecs (*Tenrec ecaudatus*) in Madagascar; the numbat (*Myrmecobius fasciatus*), a marsupial, in Australia; and the spiny anteater (*Tachyglossus aculeatus*), a monotreme, in Tasmania, Australia, and New Guinea. Hidden within this strange collection is a fascinating story of long-term phyletic survival (see chap. 16). In addition, some small rodents in both Africa and South America feed heavily on ants and termites, only one of which has been measured (*Oxymycterus roberti*). And at least one bird feeds heavily, if not exclusively, on ants and termites: the southern anteater-chat (*Myrmecocichla formicivora*) from southern Africa.

When a diverse set of unrelated species share a behavior, it is appropriate to ask whether they share a consequence derived from that similarity. For example, most large mammalian ant and termite eaters have low basal rates (fig. 5.5) compared with those expected from equation (1.4). Are the basal rates of these mammals low because they feed on ants and termites or because they belong to families that have low basal rates? How can one separate these influences? Similarly, are the BMRs of most felids and mustelids high because they are strict carnivores or because they belong to families that have high basal rates? What would it mean to argue that the low BMRs of most ant- and termite-eating mammals reflect family membership independently of food habits or other shared characteristics? But if that is the case, why do some groups within these families have higher and others lower BMRs, or is energy expenditure so unimportant that it can vary willy-nilly without any consequences? Again, if the level of energy expenditure is irrelevant, why are so many—nearly all—ant and termite eaters characterized by low mass-independent basal rates of metabolism, or to put it another way, why don't some ant and termite eaters have high BMRs, reflecting a different ancestry?

BOX 5.5. ···

Zoos and Ant Eaters

Criticism of the use of zoo animals for the study of energetics is potentially justified because the physiology of these animals may have been modified due to a lack of exercise and the consumption of an unnatural diet. However, there is no way that one can go into the field and capture tigers, elephants, giant pandas, or harpy eagles to measure their rates of metabolism. The use of zoo animals permits a greater variety of species to be included in any analysis.

Many of the ant eaters that I have measured were obtained from, or studied at, zoos, including Perth Zoo (*Myrmecobius*), San Antonio Zoo (*Tolypeutes*), Dallas Zoo (*Zaglossus*), Crandon Park Zoo (*Tachyglossus, Orycteropus*), U.S. National Zoo (*Elephantulus, Petrodromus, Oxymycteris*), and Lincoln Park Zoo (*Macroscelides, Tamandua, Cabassous, Tolypeutes, Priodontes*). *Myrmecophaga* was obtained from a dealer and after the study was donated to a zoo. Other anteaters were captured in the wild: *Cyclopes* in Amazonas, Brazil, and the Canal Zone, Panama, and *Tamandua* on Barro Colorado, Panama. With the help of Tracy Carter and Jim Shaw, the body temperatures of free-living *Myrmecophaga* were measured at Serra de Canastra National Park, Minas Gerais, Brazil, after running the animals down on foot.

But zoos are not neutral institutions. Working at the Oklahoma City Zoo was a memorable experience. I had been told by the director that I could make some measurements on an aardwolf, the only one in North America at the time. As soon as I got there, I set up the equipment to make some measurements of its rate of metabolism. I had made about eight or so measurements when it appeared that the animal had stopped eating. One of the curators and I concluded that we should leave the animal alone for a couple of days. I had seen an Indian pangolin in the zoo, so I requested permission to make measurements on it while the aardwolf recovered. I had worked on the pangolin for two or three days when a man showed up. He wanted to know what I was doing, and I explained. He said that today was my last day because I could continue on neither the aardwolf nor the pangolin. And that was that! It turned out that he was supervisor of the section of the zoo that included these animals. I went to see the director, who by then had disappeared.

I had a more amusing experience at the Tampa Zoo. I had collaborated with Charles Salisbury, the director of the zoo, in a study on the energetics of New Zealand parrots. One day as he was guiding me through the zoo, I spotted a sloth bear (*Melursus*). I made a comment that I often make: "That animal has never had its metabolism measured." He turned to me and asked, "Do you want it?" I laughed, thinking it was a joke. No, he said, you can borrow it. So, I did, taking it back to Gainesville. (The university was more tolerant of my bringing lions, tigers, jaguars, and bears into my lab than they would ever be now—or, more likely, they didn't know.) I made a series of measurements on the bear and returned it to the zoo. Charles then asked me, "Do you want another?" Therefore, I traded one sloth bear for another. The rates of metabolism of the two individuals agreed with each other.

··

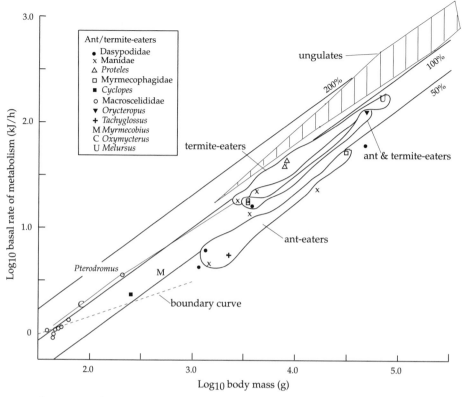

Figure 5.5. Log$_{10}$ basal rate of metabolism as a function of log$_{10}$ body mass in ant- and termite-eating mammals. A polygon for the basal rates of ungulates is included as a standard for large mammals. Also indicated is the boundary curve. (Data from McNab 2008a.)

The smallest ant and termite eaters, including elephant shrews (Macroscelididae), a rodent (*Oxymycterus*), and *Cyclopes*, scale their BMRs differently than large species in a manner that avoids entrance into torpor (see fig. 5.5). However, even though *Cyclopes* is above the boundary curve and does not enter torpor in the sense of lethargy, its temperature regulation at ambient temperatures <15°C is highly variable, as might be expected in a relatively small species (249 g) with a low basal rate (62%) (McNab 1984).

Any answer other than the pivotal importance of food habits (or other shared characteristics) is unlikely: belonging to a particular taxonomic entity may predispose species to have certain food habits, but it is likely that the characteristics of the food habits themselves influence BMR. In other words, taxonomic affiliation (i.e., phylogeny), may have an indirect impact on BMR, possibly through the probabilities of modification of physiologically dependent behaviors and, occasionally, the evolution of unique structures. Phylogeny, of course, reflects the adjustments made by a clade over time to conditions in the environment and to its behavioral repertoire. Nevertheless, there are some examples that clearly demonstrate the historical importance of phyletic affiliation (see chaps. 13 and 14).

A possible exception to the pattern of low basal rates in this heterogeneous collection of ant and termite eaters is the aardwolf. In contrast to other ant and termite eaters, it had a BMR only slightly less than the value expected from mass (95% according to Anderson et al. [1997], but 88% measured in one individual by McNab [1984]). Its higher basal rate potentially reflects its hyaena affiliation (the striped hyaena, *Hyaena hyaena*, has a basal rate equal to 88% of the expected value [McNab 2000a,c]), although another explanation is possible.

A difference exists between the basal rates of ant eaters and termite eaters. Termites are a higher-quality resource than ants because they contain a higher proportion of body fat and a lower proportion of the indigestible chitin that constitutes the exoskeleton. In fact, mammals that preferentially consume termites have basal rates that are 184% ($P < 0.0001$) of those of mammals committed to feeding on ants, but there is no difference between species that feed on a mixture of ants and termites and those that feed only on termites (McNab 1984). Thus, the aardwolf (Richardson 1987) and the sloth bear (*Melursus ursinus*), which has a BMR equal to 80% (McNab 1992a), are both specialists on termites, and they both have higher mass-independent basal rates than ant specialists (McNab 2000a). So, the higher basal rate in the aardwolf may reflect its preferential consumption of harvester termites as well as occasional rodents and bird's eggs (Kingdon 1977), a behavior not found in committed anteaters. The sloth bear has a similar diet.

The determinative factor for setting metabolism undoubtedly varies with the food. The problem with feeding on ants and termites is most obvious in large consumers, which have the greatest reductions in BMR (McNab 1984). The giant anteater (*Myrmecophaga tridactyla*), an ant-eating specialist with a mass of 30–40 kg, must take in a large volume to satisfy its energy and material needs. Even though it has a low BMR for its mass (44%), its basal rate at a mass of 30.6 kg is 1.25 MJ/d. It cannot feed on individual ants, so when it breaks open an ant nest, it consumes small particles with its tongue. Up to 50% of the intake of ant eaters that weigh >1 kg is carton from the nest or sand grains (Smithers 1971; Cooper & Skinner 1979; Bothma & Nel 1980; Greegor 1980), which of course have no nutritional value (fig. 5.6). A similar situation is found in the aardvark (*Orycteropus afer*), which has a BMR equal to 74%, which, at 48 kg, is equivalent to 2.96 MJ/d. This species feeds on a mixture of ants and termites (Willis et al. 1992), but also ingests up to 50% detritus. However, small anteaters, such as elephant shrews (40–200 g), have an indigestible intake <1% (Rathburn 1979), which accounts for their

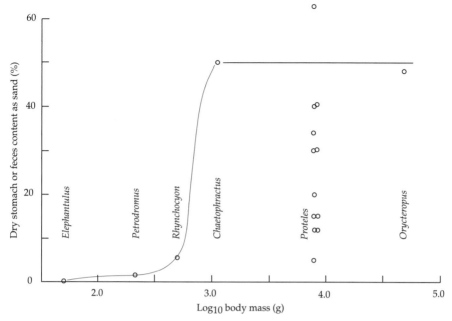

Figure 5.6. Proportion of dry stomach or feces contents consisting of sand as a function of $\log_{10}$ body mass in mammalian ant and termite eaters. (Modified from McNab 1984.)

relatively high basal rates (80–110%), as is the case in *Oxymycterus* (84 g), which has a BMR equal to 108%.

If only 40% of the mass of an ant is digestible (because its chitinous exoskeleton is indigestible), then 20% of the intake by volume by a large anteater is usable. This approximation suggests why large ant eaters cannot have high rates of metabolism: their food supply, although abundant in many tropical environments, is of a very low quality and requires extensive handling and processing times. Furthermore, ants and termites defend their nests, so most ant and termite eaters break into a nest, feed intensely for a short period, abandon the nest when the counterattack becomes too intense, and move to another nest.

Do the physiological characteristics of an ant or termite eater reflect whether it evolved its food habits from an insectivorous, carnivorous, or omnivorous ancestor or whether it retained them from an ant- or termite-eating ancestor? As noted, present evidence indicates that the energy expenditures of ant and termite eaters are more similar to one another than expected from their diverse ancestry, irrespective of whether they are monotremes, marsupials, tamanduas, pangolins, or the aardvark. This consistency counterbalances the concern about statistical significance when data from only one or two species are available (Cooper & Withers 2006). Furthermore, BMR in eutherian ant and termite eaters is reduced to near the level of mass-independent rates found in the monotreme *Tachyglossus* and the marsupial *Myrmecobius* that have similar food habits (see fig. 5.5). The differences in metabolism among monotremes, marsupials, and eutherians therefore tend to disappear when these mammals feed on low-quality resources, an observation that cannot be explained by ancestry.

A comparison of the impact of food habits on energetics in birds and in mammals

The most recent analysis of the association of food habits, when simplified into three categories, with BMR in 639 species of mammals (McNab 2008a) demonstrated that, compared with species that fed on vertebrates, fruit, nectar, bulbs, or grass, species that fed on insects, leaves, or worms or were omnivores had basal rates that were 90%, and those that fed on blood or ants and termites had basal rates that averaged 71% (see table 3.1). A similar analysis of the influence of food habits on the BMR of 533 species of birds (McNab 2009a) indicated that food habits also condensed into three categories: compared with species that fed on nuts, nectar, and pollen, species that fed on insects, seeds, aquatic vegetation, or vertebrates or were omnivorous had basal rates that were 93%, whereas those that that fed on fruit had basal rates that averaged 69% (see table 3.2). The correlation of BMR with feeding on fruit and vertebrates in birds appeared different from that found in mammals: fruit eating and vertebrate eating was associated with high basal rates in mammals, whereas vertebrate eating was associated with intermediate and fruit eating with low basal rates in birds. Why this difference?

A recent analysis was made of the basal rates of nine species of avian frugivores, including seven toucans (Rhamphastidae), one barbet (*Trachophonus darnudii*, Lybiidae), and one hornbill (*Rhyticeros plicatus*, Bucerotidae), supplemented with data on three mousebirds (Coliidae), fifteen pigeons, and nineteen pteropodid bats (McNab 2001). To this group can be added fourteen phyllostomid bats, one parrot (Psittacidae), four manakins (Pipridae), one flycatcher (*Mionectes oleaginous*, Tyrannidae), one honeyeater (*Melipotes fumigatus*, Meliphagidae), two crested berrypeckers (Paramythiidae), four bulbuls (Pycnonotidae), six birds of paradise (Paradisaeidae), five primates, five rodents, two artiodactyls, and six marsupials, all of which are frugivores. These data fall into two main groups: the eighteen passerine frugivores have significantly ($P > 0.0001$) higher mass-independent basal rates that average 1.65 times those of the collective of other frugivores, nonpasserine and mammalian, which clumped together (fig. 5.7). The one

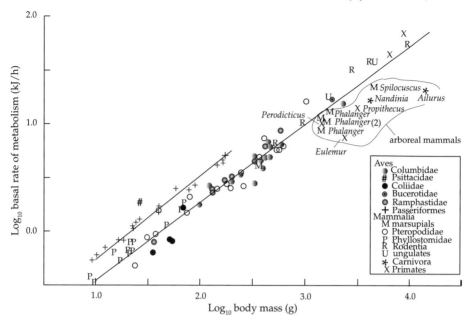

Figure 5.7. Log$_{10}$ basal rate of metabolism as a function of log$_{10}$ body mass in avian and mammalian frugivores. (Data from McNab 2008a, 2009a.)

frugivorous parrot (*Loriculus galgulus*) measured also has a high basal rate. An interesting complication occurs among mammals: some arboreal species, especially frugivorous primates, including *Eulemur* and *Propithecus*, and frugivorous or folivorous carnivorans, including *Nandinia* and *Ailurus*, have very low basal rates. Thus, the principal factor determining the BMR of avian and mammalian frugivores is body mass. Exceptions are found in the high basal rates of Passeriformes, and possibly parrots, and in the low basal rates of some arboreal mammals. The fact that fruit eating led to high mass-independent basal rates in mammals and low mass-independent rates in birds reflects the use of different mass standards: high in birds and low in mammals.

Furthermore, birds that eat vertebrates and weigh <300 g have basal rates that are low *for birds*, but have basal rates similar to those of terrestrial mammals with the same food habits (fig. 5.8). These birds include smaller *Accipiter* and *Falco*, whereas the larger falconiforms, especially those that search for prey by soaring, have appreciably lower mass-independent basal rates. This difference reflects a correlation of the basal rate in birds with pectoral muscle mass (see chap. 3). This pattern emphasizes the importance of food habits in determining BMR, with history (phylogeny) taking a minor role.

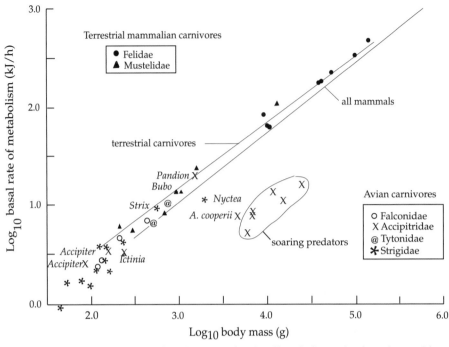

Figure 5.8. $\log_{10}$ basal rate of metabolism as a function of $\log_{10}$ body mass in avian and terrestrial mammalian carnivores. (Data from McNab 2008a, 2009a.)

The generally high mass-independent basal rates of mammalian carnivores, compared with the mammalian standard, probably reflect the ability of these animals to outrun or outswim their prey, which presumably requires a large muscle mass and an effective circulatory system, both factors leading to a high BMR. No available evidence indicates that the variation in the BMRs of carnivorous mammals is associated with body composition, except potentially in the margay.

The energy expenditures of owls are generally below the values expected from mass, as expected in sedentary species: small species have slightly low BMRs compared with mammalian carnivores, but larger species, such as *Strix occidentalis* and *Bubo virginianus*, with the exception of the snowy owl (*Nyctea scandiaca*), have basal rates equivalent to those of mammalian carnivores (see fig. 5.8). The northern hawk owl (*Surnia ulula*), which has not been measured, would be expected to have a high basal rate, since it is reputed to pursue its prey like an accipiter. Also interesting would be data from the great gray owl (*Strix nebulosa*), in part because of its large mass and northern distribution.

Nutritional factors in food affecting BMR

All foods are not created equal. They differ in many ways that profoundly influence the energetics of endotherms, especially because most endotherms are committed to an energetically extravagant lifestyle. Foods differ in terms of their energy content, carbohydrate/protein ratios, secondary compounds, abundance, and seasonality, all of which may affect the energy expenditure of their consumers. Cruz-Neto and Bozinovic (2004) recently examined the potential relationship between the BMRs of species and the chemical composition of the foods that they consume. They emphasized the value of intraspecific studies, which eliminate many of the complexities found in interspecific comparisons. The range of food habits found intraspecifically, however, is very small. These authors concluded that we are far from understanding the connections that might exist between food makeup and energy expenditure and, to emphasize that point, cited a series of ambiguous or conflicting results. However, a preoccupation with "the" food-habit hypothesis is misplaced. Evidence clearly indicates a correlation of BMR in some groups with the foods consumed, but it does not at all address the basis of such correlations, which is undoubtedly exceedingly complicated, considering the range of species of birds and mammals studied and of foods consumed. The low BMRs of bats that feed on figs (McNab 2003b), in contrast to species that feed on *Piper*, *Vismia*, or *Solanum*, may reflect the lower protein and lipid content of figs (Fleming 1986), but the real cause of the low BMRs of these species may be quite different. (To delve deeper into this question, see Karasov and Martinez del Rio 2007.)

CHAPTER SIX. *Life in the Cold*

Endotherms respond to a permanently or seasonally cold environment in a variety of ways. They may adjust basal rate, thermal conductance, and body mass, rescale basal rate at small masses, or avoid cold conditions, either through migration or by selective entrance into torpor (see chap. 8).

Adjustment of basal rate

Adjustment to life in the cold occurs in a variety of circumstances, including latitudinally cold climates, altitudinally cold climates, seasonally cold climates, and aquatic environments.

A latitudinally cold climate

As we saw in chapter 2, Per Scholander, Laurence Irving, and colleagues, in three fundamental articles (1950a,b,c), compared the BMRs, thermal insulation, and body temperatures of terrestrial and aquatic birds and mammals from Panama and Alaska to determine whether climate had an impact on the energetics of endotherms. These authors concluded that body mass alone determined BMR, although they noted that "small adaptive changes cannot, of course, be detected by such interspecific comparisons." But, they continued, "intraspecific evidence from observations on man . . . have failed to reveal any certain racial metabolic or body temperature adaptations to cold climates (Dubois 1936). Man and animal alike do it all with insulation." Interestingly, these authors limited their conclusions to terrestrial mammals, even though they presented data on birds and aquatic mammals as well.

Contrary to their conclusions on terrestrial mammals, their data appeared to show that Arctic birds and aquatic mammals had higher basal rates than tropical species. Water is a "colder" environment due to its high thermal conductivity and heat capacity. A justifiable concern, however, is that none of the Arctic species were closely related to the tropical species, so many of the differences found between these climatic groups may have reflected differences in behavior, ecology, or phylogeny as well as climate. These authors' implication that the differences in BMR between Panamanian and Alaskan species might reflect differences in "phylogeny" was to resurface in the 1990s (see chap. 2).

One way to resolve this ambiguity is to compare populations of the same species, or of closely related species, that live in different climates so that differences among species in evolutionary history, ecology, and behavior are minimized. This was first accomplished in deer mice (*Peromyscus*) (McNab & Morrison 1963). Since that time, many biologists have examined the question of a climatic adjustment in energy expenditure.

Studies have collectively indicated that species and genera of birds and mammals have BMRs that appear to reflect the climatic conditions they face: populations and species living in cold climates have higher BMRs than populations and species living in warm-climate or low-altitude environments (Chaffee & Roberts 1971). The many examples of adjustment of BMR to a cold climate include house sparrows (*Passer domesticus*): individuals that live in a cool, temperate environment have higher basal rates than individuals that live in a seasonally hot, humid environment (Hudson & Kinzey 1966). Boreal populations in two species of chipmunks (*Tamias*) from Northwest Territories and Alberta had higher BMRs than alpine populations from Alberta and Colorado (Jones & Wang 1976), and all of these populations had high basal rates compared with the general mammalian curve. Basal rates in terrestrial and marine birds increase with latitude (Weathers 1979; Ellis 1985), as they do in falconiforms (Wasser 1986), although in the latter case, body composition and predatory style, as well as long-distance movement, complicate matters (see chaps. 3 and 8). Resident populations of the stonechat (*Saxicola torquata*) from Kenya have lower BMRs than sedentary populations from Ireland (Klaassen 1995; Wikelsi et al. 2003a). (Here we return to the possibility that the high BMRs of migratory populations reflect body composition and activity level.) Polar and temperate mammals have higher BMRs than temperate-tropical and tropical species (McNab 2008a), and temperate passerines have higher basal rates than tropical species (Wiersma et al. 2007; McNab 2009a), although this generality must be accepted with caution because tropical environments are not uniform and the diversity of the tropical species studied is greatly restricted. Life in a cold climate, therefore, appears to be an important factor influencing basal rate of metabolism: a change in insulation, contrary to the conclusion of Scholander et al., is not the only response.

Another potential approach to the influence of climate on the energy expenditure of mammals is to describe it in relation to zoogeography (Lovegrove 2000). Mammalian BMR correlates with zoogeographic regions:

mean BMRs are low in the Afrotropical, Australasian, Indomalayan, and Neotropical regions, whereas they are high in the Nearctic and Palearctic regions. These findings, of course, also represent correlations with latitude and longitude, components of climate. However, the correlation with zoogeographic regions also includes an element of history; in other words, the geographic distribution of mammalian taxa is not uniform. For example, the Afrosoricia are limited to Africa and Madagascar; the Pilosa and Cingulata are Neotropical, with a small contingent marginally present in the Nearctic region; the Metatheria occur in the Neotropical, Australasian, and marginally in the Nearctic regions; and the Monotremata are found today only in the Australasian region. The differential geographic distribution of taxa has an impact on the frequency distributions of mass-independent basal rates. Thus, the presence of monotremes and the accumulation of marsupials in Australia and New Guinea lead to a low mean basal rate in the Australasian region. And the diversity of marsupials and pilosans in the Neotropics contributes to its low mean basal rate.

An altitudinally cold climate

Little effort has gone into examining the possibility that the energetics of endotherms respond to the decrease in ambient temperature with an increase in altitude. Limited evidence shows that frugivorous and nectarivorous bats that live in tropical mountains have higher basal rates than lowland populations and species (Bonaccorso & McNab 1997; McNab & Bonaccorso 2001; Soriano et al. 2002; McNab 2003b), a pattern seen in eutherians generally (see chap. 3). However, marsupials have a reduced diversity at high altitudes, as they do in cold-temperate climates (see chap. 13), so it is not surprising that marsupials at intermediate altitudes in Papua New Guinea have lower BMRs than their low-altitude relatives (McNab 2008b).

Birds of paradise collectively occur at altitudes from sea level to 3,500 m. Species that are limited to altitudes $\leq 1,000$ m in New Guinea have basal rates—as always, corrected for body mass—that are 91% of those of species found at higher altitudes (McNab 2003c, 2005a). As seen in chapter 5, the basal rates of birds of paradise are also correlated with their food habits (fig. 6.1). These factors may be included in the following equation:

$$\text{BMR (kJ/h)} = 167.5 \, (F \cdot A) \, g^{0.879}, \qquad (6.1)$$

where F is the dimensionless coefficient for food habits and A is the coefficient for altitude. F equals 1.00 for omnivores, 0.91 for insectivores, and 0.76 for frugivores. Body mass alone accounts for 91.7% of the variation in these birds' BMR, whereas equation (6.1) accounts for 99.0% of the variation. For example, the superb bird of paradise (*Lophorina superba*) has a BMR equal

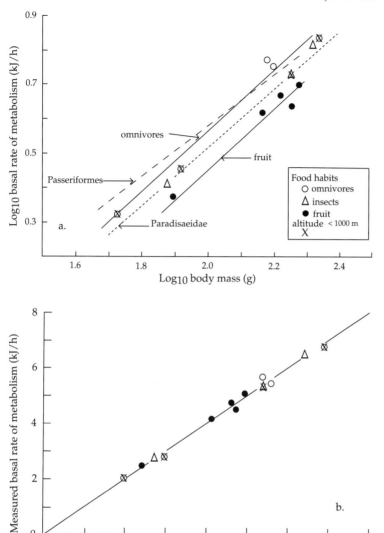

Figure 6.1. (A) $\log_{10}$ basal rate of metabolism as a function of $\log_{10}$ body mass in birds of paradise (Paradisaeidae) in relation to food habits and altitudinal distribution. (B) Measured basal rate of metabolism in birds of paradise as a function of the basal rate calculated from equation (6.1). (A modified from McNab 2005a.)

to 92% of the value expected from the all-bird curve, pulled down by its insectivorous habits, but raised by its distribution above 1,000 m.

Why some species have higher BMRs at higher altitudes is unclear, unless high basal rates are a response to lower ambient temperatures. That probably is unlikely because in tropical locations such as New Guinea, altitudes

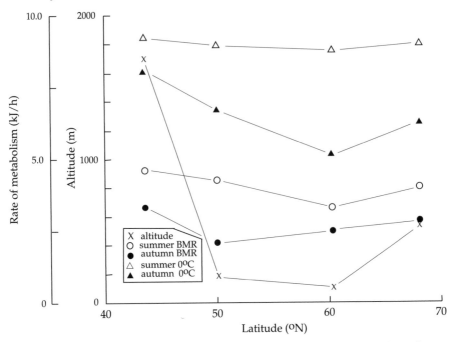

Figure 6.2. Rate of metabolism and altitudinal distribution as a function of latitude and time of year in *Myodes glareolus*. (Data from Aalto et al. 1993.)

up to 2,000–3,000 m are not as cold as a mid-latitude temperate climate. The absence of an increase of BMR in marsupials at high altitudes and latitudes may reflect their inability to increase energy expenditure generally (see chap. 3). A limited altitudinal distribution of available food resources may also restrict marsupials (see chap. 12).

A climatic impact associated with latitude may be avoided if a species adjusts its local altitudinal distributions to ensure that it faces similar climatic conditions, such as temperature and rainfall, independently of latitude. For example, the bank vole (*Myodes* [= *Clethrionomys*] *glareolus*) showed no change in basal rate over a 3,000 km range in latitude because it increased its altitude at southern latitudes, shifting from 110 m in Finland to 1,700 m in Bulgaria (fig. 6.2) (Aalto et al. 1993). A similar distribution pattern occurs in other arvicolines in North America, including *Synaptomys cooperi*, *Phenacomys intermedius*, and *Microtus richardsoni*.

A seasonally cold climate

Season has also been shown to influence BMR in mammals and birds, especially in species that live in temperate and polar environments. This influence was clearly seen by Kendeigh et al. (1977), who demonstrated in a summary of data collected on Russian birds that winter-acclimatized individuals generally had higher basal rates than fall- or summer-acclimatized individuals of the same species. A similar pattern is seen in mammals (Hart 1971), although some species, such as deer (*Odocoileus virginianus* and *Capreolus capreolus*), reduce energy expenditure in winter (Silver et al. 1969; Weiner 1977), possibly reflecting a switch in diet from grass to browse (Macari et al. 1983). As seen in these deer, however, geographic patterns in energy expenditure are complicated by interactions among many factors, in this case climate, season, and food quality.

An aquatic life

Another complication is seen in aquatic mammals. Species that live in cold water have very high basal rates of metabolism (see fig. 5.3), especially those that are carnivorous (see chap. 5). However, even three cold-water rodents, the beaver (*Castor* [129%]), muskrat (*Ondatra* [126%]), and coypu (*Myocastor* [208%]), have high basal rates (Arends & McNab 2001). A large tropical aquatic rodent, the capybara (*Hydrochaeris hydrochaeris*), has a rather high basal rate (123%). The platypus (*Ornithorhynchus anatinus*) has the highest basal rate among monotremes (76%), whereas the yapok (*Chironectes minimus*), an aquatic marsupial, has an intermediate basal rate (77%), but it is not found in cold water.

Life in cold water, therefore, is usually associated with a high basal rate, possibly in compensation for increased heat loss. This pattern is seen irrespective of food habits, although carnivory appears responsible for the differentially high basal rates of carnivorous species (i.e., the shift from ca. 125%–200% in aquatic rodents to 170%–350% in aquatic carnivores). The very high basal rates of porpoises may also reflect a body composition associated with high activity levels. Data from tropical monk seals (*Monachus*) or Amazonian porpoises (*Inia*, *Sotalia*) might permit us to separate the influences of climate (water temperature) and of food habits (carnivory) on basal rate. These species may well have high basal rates, but probably not as high as those of their cold-water relatives. Of special interest is the observation that the basal rate of the Amazonian manatee (*Trichechus inunguis*) is very low (50%), which undoubtedly reflects in part its tropical distribution, sedentary lifestyle, and herbivorous food habits. The data on aquatic mammals clearly demonstrate the difficulty of assigning the degree to which various factors determine the residual variation in mammalian BMR because we depend on correlations for the analyses, which are therefore fraught with difficulties.

The value of a high BMR

The usual explanation for high BMRs in cold-climate endotherms is that they compensate for the greater temperature differentials maintained in aquatic, polar, temperate, or montane environments (McNab 2008a, 2009a). A possible basis for this explanation can be seen in figures 1.1 and 1.2B, in which an increase in BMR leads to a decrease in the lower limit of thermoneutrality when mass and conductance remain the same.

The rationale for the increase in BMR may actually be much more complicated. First of all, an increase in BMR might lead to a greater increase in the daily energy expenditure than could be saved by any minor decrease in the lower limit of thermoneutrality. Second, the increase in BMR with cold acclimatization may actually be associated with a change in some correlate of BMR, such as the maximal rate of metabolism (Lindström & Kvist 1995), which would permit an increase in the maximal ΔT that can be maintained and therefore an increased tolerance of low ambient temperatures. Yet a reexamination of the data on passerine birds (B. K. McNab, personal observation) does not show a significant correlation ($P = 0.11$) of the residual variation in maximal rate with the residual variation in basal rate. Another possible explanation might be that an increase in BMR permits an increase in fecundity that balances an increase in mortality in a cold environment (McNab 1980b), at least in mammals, as long as the increase in metabolism can be sustained (see chap. 14). Or, all these factors may contribute to selection for a high basal rate in permanently or seasonally cold climates.

The adjustment of thermal conductance

As noted, there have been few studies of seasonal adjustment of thermal conductance (and insulation), except to note that species active during cold winters either show an appreciable decrease in thermal conductance or, if small, little seasonal change. Insulation is normally increased through an increase in the density or thickness of the fur coat. A small mass limits the thickness of the coat that can be carried, but insulation can be potentially increased in small species through an increase in the density of the coat. For example, along an altitudinal gradient from 1,524 to 3,353 m in Colorado, *Peromyscus maniculatus*, which weighed 15.6 to 19.5 g, showed a 30% increase in fur length (from 7.2 to 9.3 mm) at an altitude of 2,438 m, with no further increase at higher altitudes, whereas fur density increased up to 3,353 m (Wasserman & Nash 1979). Whether a limit on the thickness of insulation applies to birds is unclear.

One of the most interesting small Arctic mammals with regard to behavior and thermal conductance is the varied lemming (*Dicrostonyx groen-landicus*), which, unlike the brown lemming (*Lemmus trimucronatus*), spends much of its life in winter above the snow layer (which accounts for its cryptic white winter coat). Its thick fur coat is exceedingly dense and has a very low thermal conductance (McNab 1992b). Most small mammals that live in temperate or Arctic regions, like the brown lemming, avoid the coldest temperatures and wind chill by seeking warm microclimates, usually under leaf litter or in runways under the snow. This behavior reduces their energy expenditure compared with what it would be if they were exposed to the coldest microclimates on bare ground or above the snow, and thereby lessens their need for increased insulation. Indeed, thermal conductance in winter *Dicrostonyx* was 42%–54% of the value expected from a mass of 60 g, whereas it was 64%–78%, at 45 g, in *Lemmus* (McNab 1992b); in other words, conductance was lowest in the lemming most exposed to the coldest temperatures.

Arctic ptarmigan (*Lagopus* spp.), which have an intermediate size, seek shelter at night in snow burrows (Korhonen 1980), which probably raises nighttime ambient temperatures for these birds to slightly below 0°C. Small birds in cold-temperate winters avoid the coldest temperatures overnight by using shelters to reduce heat loss.

A change in body mass

Some mammals respond to changes in climate or season with a change in body mass. Several situations in which a change in body mass occurs have been the basis of ecological "rules." These patterns have been recently revisited (McNab 2010).

Bergmann's rule

The classic example of a climatic rule is Bergmann's rule, which states that mammals tend to be larger in colder climates. Carl Bergmann first described this pattern in 1847, twelve years before Darwin's *Origin of Species*, with the implication, if not the direct conclusion, that the change in mass was an adaptation to climate! (Was this the concept of evolution before Darwin, or just god's—the engineer's—wisdom in animal design?) Bergmann's analysis was based on interspecific comparisons, although Ernst Mayr (1956) argued that the "rule" should be restricted to intraspecific comparisons. Species differences, however, are usually the evolutionary extension of intraspecific populational differences. Bergmann's rule selectively applies interspecifically, as we shall see, but the caution of Mayr is well taken, considering the potential complications in interspecific studies, as was seen in the Scholander-Irving studies.

Mammals have often been surveyed to determine the extent to which they conform to Bergmann's rule. There is little agreement among these surveys. For example, most North American mammal species with a wide latitudinal distribution either did not show a correlation of size with latitude or did so over only a limited latitudinal range; one-third of the species showed a decrease in size with latitude (McNab 1971). According to Dayan et al. (1991), only a minority of carnivore species conform worldwide to Bergmann's rule, yet Meiri et al. (2004) concluded that 50% of carnivores conform. The preoccupation with whether 50% or more of species conform to a pattern is misplaced, except as one is preoccupied with the term "rule." If a species follows Bergmann's rule, or contravenes it, the size cline is worthy of study because it tells us something of the biology of this species, irrespective of whether it conforms to a majority or a minority trend.

Fewer studies have examined conformation to Bergmann's rule in birds than in mammals. Several early investigations indicated that birds indeed conformed (Rensch 1936; James 1970), although Zink and Remsen (1986) found that a minority followed the rule, but they showed that it was most marked in sedentary species, as did Rensch. Two recent reviews of this relationship have occurred. Ashton (2002) found that 76% of examined species conformed to the rule, in terms of both latitude and ambient temperature. A study by Meiri and Dayan (2003) demonstrated that 72% of examined birds conformed to the rule. Again, sedentary species more often were larger at higher latitudes, as might be expected from the propensity of high-latitude migratory species to avoid the lowest seasonal temperatures. One of the difficulties in the bird studies is that insufficient data on body mass are available, so a common measure of avian size, wing length, is used, which raises the problem that factors other than body size influence flight. The same problem is found in mammals, but the usual measure of body size is head-body length, which is a better measure of body mass than wing length is in birds.

In an extensive survey of previous studies, Ashton et al. (2000) argued that 71% of examined species of mammals showed a positive correlation of body mass with latitude and that this correlation was principally due to an inverse correlation of body size with ambient temperature. These authors did not approve of analyses that subdivided latitudinal ranges, but they, and the people they quoted, usually forced a linear relationship between size and latitude, irrespective of any local geographic trends. Ashton and colleagues also raised the question of whether smaller species might be more likely to follow Bergmann's rule, but concluded that was not the case. They further concluded that their analysis did not support the conclusion that heat conservation was the basis of the conformation of mammals to Bergmann's rule, although they made no suggestion as to its actual basis. Meiri and Dayan (2003) came to the conclusion that 65% of examined mammals conformed to the rule.

The commonest interpretation of an increase in mass in a cold climate is that it leads to a decrease in the rate of metabolism. The basis of this interpretation is that total rates are proportional to mass raised to a power $b <$ 1.00, such as 0.72 (see eqs. [1.4] and [1.5]). As a result, mass-specific basal rates are proportional to $m^{0.72}/m^{1.00} = m^{-0.28}$ and therefore decrease with an increase in mass. However, mass-specific rates are not ecologically relevant (and in a sense do not even exist, except as an intellectual abstraction); total rates are relevant, and they unambiguously increase with body mass (McNab 1999) (see chap. 1). So, what is the value of having a larger size in a cold climate, when that occurs, given the associated increase in total rate of metabolism?

One potential value of a larger size is that larger individuals can tolerate harsh conditions better than smaller individuals because starvation time increases with mass. This interpretation has been suggested several times (Boyce 1978, 1979; Calder 1984; Lindstedt & Boyce 1985; Millar & Hickling 1990; Mugaas & Seidensticker 1993; Sand et al. 1995). Its rationale is as follows: if body energy stores are proportional to mass (i.e., $m^{1.00}$) or greater, and if individuals expend energy at rates that are proportional to about $m^{0.72}$ (see eq. [1.4]), then starvation time is proportional to $m^{1.00}/m^{0.72} = m^{0.28}$ (i.e., the inverse of mass-specific basal rates). An individual whose weight is 1.25 times that of another, then, should be able to survive without food for $(1.25)^{0.28} = 1.06$ times as long as (i.e., 6% longer than) the smaller individual, all else being equal.

That difference is small in small species, but it may represent a significant increase in large species. In fact, most of the studies that have advocated the importance of starvation time have dealt with species of a medium (*Procyon lotor, Ondatra zibethicus*) or large (*Alces*) size. The only application of the starvation time argument to small species has been by Fleischer and Johnston (1982) and Murphy (1985) to *Passer domesticus*. The rationale for an increased starvation time in small species was questioned by Dunbrack and Ramsay (1993), who suggested that other survival strategies would be more likely, including food storage, selection of microclimates, or daily or seasonal torpor.

Rosenzweig (1968) proposed that an increase in consumer size with latitude reflects an increase in plant and community production with latitude, a view endorsed by Guthrie (1984) and Geist (1987). Huston and Wolverton (2009), in fact, showed that the relevant terrestrial primary production increases with latitude and reaches its maximum at latitudes between 30° and

50°, beyond which it decreases. Wolverton et al. (2009) applied this concept to the conformation to Bergmann's rule in white-tailed deer (*Odocoileus virginianus*). As a refinement, Langvatn and Albon (1986) suggested that a latitudinal increase in the mass of the red deer (*Cervus elaphus*) in Norway reflected a latitudinal increase in foliage quality. The reduced size of ungulates in the high Arctic, as is seen in barren-ground caribou (*Rangifer tarandus*) and musk oxen (*Ovibos moschatus*), then, reflects the limited quantity and quality of food resources on the tundra (Geist 1987). A dependence of herbivore body size on primary production, reflecting variations in rainfall and temperature, is also seen in *Peromyscus* in North America (McNab & Morrison 1963; Mueller & Diamond 2001), sifakas (*Propithecus*) in Madagascar (Lehman et al. 2005), and burrowing rodents of the genus *Ctenomys* in South America (Medina et al. 2007).

A few examples of Bergmann's rule appeared to reflect a latitudinal limit on the distribution of the larger species in a competitive set of predators, beyond which the smaller species were able to increase in size because larger prey were now available to them (McNab 1971). One of the examples of this pattern was seen in some North American weasels (fig. 6.3). Three species occur in eastern North America, including, from the largest to the smallest, the long-tailed weasel (*Mustela frenata*), short-tailed weasel (*M. erminea*), and least weasel (*M. nivalis*). *Mustela frenata* has a head-body length that is more or less independent of latitude between 9° and 54° N (with a slight increase at lower latitudes). It is an aboveground predator that feeds primarily on birds and rabbits, but is limited in its northern distribution by snow depth. The short-tailed weasel is a vole predator that searches for its prey beneath the snow in winter (Simms 1979; Erlinge 1987). It has a size that is independent of latitude between 42° and 55° N, but shows an appreciable

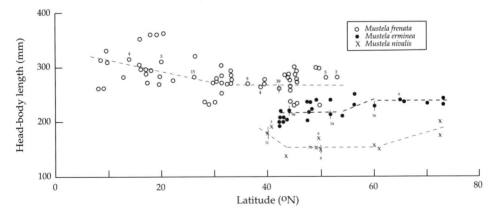

Figure 6.3. Head-body length in three species of weasels as a function of latitude. (Modified from McNab 1971.)

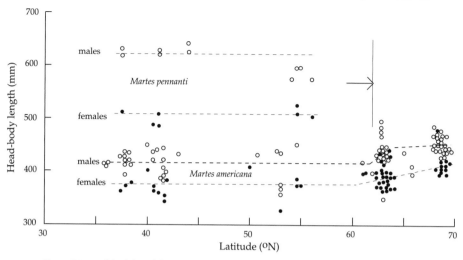

Figure 6.4. Head-body length in two species of arboreal mustelids, the fisher (*Martes pennanti*) and the pine marten (*M. americana*), as a function of latitude. (Modified from McNab 1971.)

increase in length at higher latitudes—that is, at latitudes beyond the northern distributional limit of *M. frenata*. At these higher latitudes, *M. erminea* encounters larger vole prey, the lemmings *Lemmus* and *Dicrostonyx* (Ralls & Harvey 1985). The least weasel has a length that is independent of latitude between 43° and 60° N, but it is larger at 72° N. Length and mass in smaller weasels therefore appear to reflect the presence or absence of larger weasels and the size of available prey.

A similar pattern was seen in a pair of larger, semiarboreal mustelids, the fisher (*Martes pennanti*) and marten (*M. americana*), in which the larger fisher has a size that is independent of latitude between 37° and 56° and probably 62° N, whereas the smaller marten has a size independent of latitude between 36° and 62° N, above which it shows an increase in size (fig. 6.4). However, another pair of mustelids, the larger otter (*Lutra canadensis*) and the smaller mink (*Mustela vison*), both semiaquatic species, have congruent distributions from 26° to 64° N, and no correlation of size with latitude occurs in either species. The factor influencing the presence or absence of a change in size in mustelids appears to be the division of prey by size, and it was the smaller species that got larger beyond the geographic limit of distribution of its larger competitor.

Direct evidence indicates that division of food particle sizes influences the sizes of some predators as a function of latitude, most clearly in the two largest American cats, the puma (*Puma concolor*) and jaguar (*Panthera onca*) (fig. 6.5A). The puma is smallest between 28° N and 32° S, which covers the 60° latitudinal range of the larger jaguar (McNab 1971). At higher

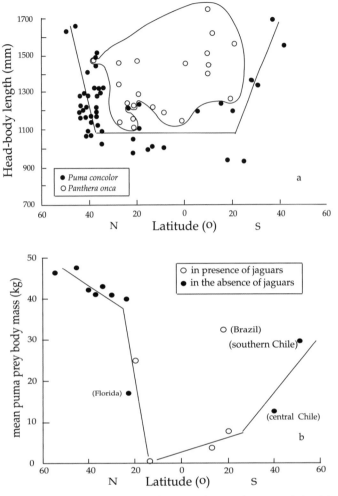

Figure 6.5. (A) Head-body length in two large cats, the puma (*Puma concolor*) and the jaguar (*Panthera onca*), as a function of latitude. (B) Mean puma prey size as a function of latitude in the presence and absence of jaguars. (A, data from McNab 1971; B, data from Iriarte et al. 1990 and McNab 2010.)

latitudes in both hemispheres, the puma attains a length that is equal to that of the absent jaguar; in other words the puma conforms to Bergmann's rule twice.

No direct evidence of division of prey sizes by these two felids was origi-nally available, but later Iriarte et al. (1990) demonstrated that the mean mass of puma prey correlated with puma length and that both were larg-est in the absence of the jaguar (fig. 6.5B). In North America beyond the geographic limit of the distribution of jaguars, puma prey and pumas both increased greatly in size (fig. 6.6). Here was evidence that the increased size of the puma reflected the absence of the competitively superior jaguar and the necessity of dividing the food supply by size, as happens where they coexist. The size of available prey is responsible for the size of the puma: the puma is small in Florida, despite the absence of the jaguar, as is to be expected from the small size of its most abundant prey, white-tailed deer, on this peninsula (Maehr et al. 1990); a similar situation is present in Chile (Yáñez et al. 1986). On the other hand, in the Pantanal of Brazil, where both species feed on livestock, the puma has a large mean prey mass in the presence of the jaguar.

Recent studies have further supported the concept that geographic variation in the size of predatory mammals reflects the size of available prey. For example, among North American canids, the smaller species, including various foxes and the coyote (*Canis latrans*), have body sizes that did not vary with latitude, as might be expected in the presence of the larger wolf (*C. lupus*), but the wolf showed latitudinal variation in size (McNab 1971). Wolf body length increased between 16° and 40° N, above which size progressively decreased to 80° N (fig. 6.7). Geist (1987) argued that the size of wolves depends on the size of their prey. The wolf's principal

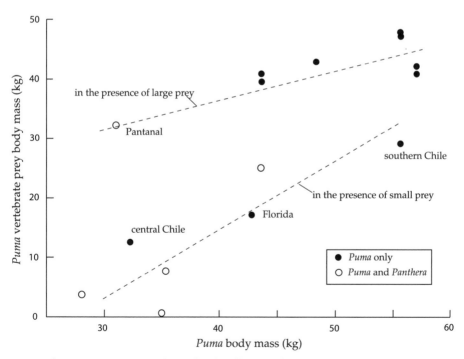

Figure 6.6. Mean puma prey size as a function of locality and puma body mass in the presence and absence of jaguars. (Data from Yáñez et al. 1986, Iriarte et al. 1990, and Maehr et al. 1990.)

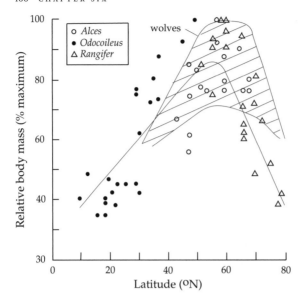

Figure 6.7. Relative body mass of moose (*Alces americana*), deer (*Odocoileus hemionus*), elk (*Cervus canadensis*), caribou (*Rangifer arcticus*), and wolves (*Canis lupus*) as a function of latitude. (Modified from Geist 1987.)

prey are ungulates, which showed a collective increase in mass from about 10° to 50°–60° N as wolves shifted from smaller prey (deer [*Odocoileus*]) at lower latitudes to larger prey (elk [*Cervus canadensis*] and moose [*Alces americana*]) at higher latitudes. A decrease in wolf size occurred at the highest latitudes, where they switched to smaller prey (caribou and musk ox) (see fig. 6.7). Therefore, the latitudinal variation in body size in the top canid reflects the size of its available prey. Similarly, the body size of foxes in Scotland is larger in areas where they prey on hares and rabbits and smaller where they feed on voles; these differences in the available prey reflect conditions in the environment (Kolb 1978). Whether food abundance is a basis for a correlation of avian body size with latitude is unclear, but at least any increase in mass requires an increase in food intake.

A correlation of body size with prey size also exists among the smallest mammalian predators. Soricine shrews in Eurasia decrease in size with latitude (Aitchison 1987; Hanski & Kaikusalo 1989; Churchfield et al. 1999; Ochocińska & Taylor 2003). Large species have per capita food consumption that is about twice that of small shrews. Large species feed on earthworms and isopods, which are not found in Arctic or alpine areas. Small species feed principally on arthropods. Large shrews are thus restricted to lower latitudes and altitudes because of their dependence on larger prey. Two of the smallest shrews in Eurasia, *Sorex minimus* and *S. tundrensis*, are found above the Arctic Circle, where no large shrews are found. The smallest (2.1 g) shrew in North America, *S. hoyi*, has the northernmost distribution.

Dehnel's phenomenon

Another size cline is Dehnel's phenomenon. If an endotherm needs to reduce its energy expenditure in a cold climate without giving up temperature
regulation, the principal way that can be accomplished is by reducing body
size because, again, the total rate of metabolism is what is ecologically and
evolutionarily relevant. In 1949, August Dehnel described a tendency for
some shrews to reduce body mass in the fall in apparent anticipation of
winter conditions. This reduction was not simply a reduction in body fat,
but involved a reduction in the size of the skeleton and brain. Since then,
this phenomenon has been found in many shrews of the genus *Sorex* (Pucek
1963, 1964; Mezhzherin 1964; Mezhzherin & Melnikova 1966; Hyvärinen
1984; Genoud 1988; Merritt & Zegers 2002) and in some arvicoline rodents,
including *Microtus* (Iverson & Turner 1974) and *Myodes* (Merritt & Zegers
1991). What is most interesting about this set of species is that they are also
committed to continuous endothermy (i.e., they do not enter torpor).

This pattern suggests that small endotherms facing a cold winter must
reduce energy expenditure, but their approach is represented by a choice of
lifestyle: either use daily or seasonal torpor, which reduces energy expenditure through a reduction in the regulated body temperature, or retain a
commitment to endothermy and reduce energy expenditure by reducing
body mass. What factors determine which alternative is used?

At first glance, the choice appears to be phylogenetic in the sense that
those small mammals that "choose" a Dehnel solution are arvicoline rodents and soricine shrews. But the "choice" may be more ecological in that
it reflects food habits: species that reduce body mass consume soil invertebrates and leafy vegetation, whereas the small rodents that use torpor include cricetine rodents, such as *Peromyscus* and zapodid rodents, that feed
on seeds and berries. Crocidurine shrews of the genera *Suncus* and *Crocidura*, in contrast to soricine shrews, use daily torpor. They are principally
tropical in distribution, only marginally successful in cold-temperate environments, and do not enter polar environments. They do not conform to
Dehnel's rule, which raises the question, why not? This phenomenon needs
much more study.

A resource rule

Bergmann's rule and Dehnel's phenomenon both reflect the availability of
resources, which is also true of the "island rule" (Foster 1964). This rule
describes the response of continental mammals to conditions encountered
on islands: a decrease in the mass of large species and an increase in the

mass of small species (see chap. 9). A semi-rule is that many species or species groups tend to be smaller in desert environments (see chap. 7). Meiri et al. (2008b) suggested that the propensity of some large mammals to be smaller on the large island of Borneo than on smaller islands results from low primary production on that island, possibly related to its soil, which is characterized by a low nutrient content.

These "rules" all share a dependence on the primary production in a region, and consequently on the availability of resources, which in turn reflects conditions in the environment, including temperature, rainfall, and the necessity of sharing resources with other species. This unified pattern can be called the "resource rule," thereby reflecting the pivotal importance of resource availability (McNab 2010). A consequence of this rule is that consumers become smaller with a reduction of resources, as in the case of herbivores and predators in regions of low plant productivity, which includes the Arctic plain, desert environments, and islands in the case of large continental species. This rule has also been applied to dinosaurs: the largest sauropod and theropod dinosaurs got 6–8 times larger than ecologically equivalent terrestrial mammals because, it was argued (McNab 2009c), they expended energy at rates that were about 22% those of mammals of the same mass (see chap. 4), but never got as large as the biggest marine mammals, which live in a more resource-rich environment.

A recent article by Gardner et al. (2011) suggested that one response to global warming might be a decrease in body size in both ectotherms and endotherms. These authors accumulated data from 32 studies, 13 of which showed a decrease in size, although 16 of the studies examined had one species. They concluded that two factors might be responsible for a decrease in body size: ambient temperature per se and the availability of resources, the latter conclusion being similar to the "resource rule." If the resource rule were to apply here, the implication would be that the decrease in size reflects a decrease in the abundance or quality of resources. That pattern, however, would not be geographically or environmentally uniform because it would depend on the future distributions of water and temperature. Thus, a tropical rainforest might develop in Florida and a desert in midwestern North America, but both would be associated with an increase in environmental temperature.

Rescaling of BMR

As we have seen (in chap. 4), small mammals and birds cannot continue a commitment to continuous endothermy at small masses without modifying the scaling of BMR that is seen at large masses. If large-species scaling continues to masses that are less than about 45 g, these smaller species use torpor as a response to low ambient temperatures (see chap. 8). That is why small mammals that do not use torpor, such as soricine shrews and arvicoline rodents, have such high basal rates (reflected in the boundary curve). Most small mammals have intermediate basal rates, and many of them use torpor to varying degrees, as is the case in *Peromyscus*, depending on the availability of resources and their body size. Furthermore, the generally higher basal rates in small birds than in small mammals accounts for their less prevalent use of torpor (see chap. 4). Indeed, small mammals that do not enter torpor have basal rates that are essentially equal to those of small birds (see fig. 4.8). When seasonally harsh conditions appear, most birds migrate to avoid them, whereas small mammals generally cannot evade such conditions; thus, depending on their food habits, many species use daily or seasonal torpor. This pattern applies even to temperate bats, most of which evade the shortage of food (flying insects) in winter by hibernating in local caves. These bats do not conform to the boundary curve. Insectivorous bats of the genera *Lasiurus*, *Nycticeius*, and *Nyctalus*, however, migrate from cold-temperate environments to spend winter in warm climates and, at least in *Lasiurus* and *Nycticeius*, are more likely to thermoregulate at intermediate to low ambient temperatures than are temperate *Myotis* (Genoud 1993).

CHAPTER SEVEN. *Life in Hot Dry and Warm Moist Environments*

The adaptation of endotherms to life in warm to hot environments has attracted less attention than has adaptation to the cool to cold environments typical of Europe and North America, where most of the comparative studies of avian and mammalian energetics have occurred. Warm environments can be broken down into two extreme conditions, dry and wet, which require radically different responses.

Life in a hot, dry environment

Many biologists have examined adjustment of BMR in relation to a desert existence because much work has been done in Australia, Israel, South Africa, and the southwestern United States.

Mammals

The first tentative answer to the question of climatic adaptation, after the denial of such a possibility by Scholander et al. (1950c; see chap. 2), was found in populations and species of deer mice (*Peromyscus*), in which it was shown that a desert existence was associated with low basal rates (McNab & Morrison 1963). This study included five species. Three of these species lived in a mesic environment at Berkeley, California (*P. californicus*, *P. truei*, and *P. maniculatus*); one of the three (*P. californicus*) was also collected in the semi-mesic San Gabriel Mountains and the other two in the dry Spring and Charleston Mountains of Nevada at 1,800 to 2,100 meters. The other two species (*P. eremicus* and *P. crinitus*) were trapped in desert lowlands near Las Vegas, Nevada (see box 7.1); additional *P. eremicus* were collected from New Mexico and additional *P. crinitus* from Utah. This comparison of intraspecific populations of closely related species avoided some of the phylogenetic deficiencies in the Scholander-Irving study.

The BMR in four of the five species and eight of the ten populations,

BOX 7.1. ···

A Study of Desert Adaptations in *Peromyscus*

As an undergraduate at Oregon State University (then College), I became interested in the question of climatic adaptation after reading the Scholander-Irving articles published in 1950. Climatic adaptation could be better examined by comparing closely related species, thereby minimizing the phylogenetic factors that complicate all analyses. An appropriate study was to compare various species of the genus *Peromyscus* that live in wet and dry environments. That led me to choose this project upon moving to the University of California (Berkeley) for graduate work with Oliver P. Pearson. However, because "Payne" Pearson decided to retire prematurely, I transferred to the University of Wisconsin to work with his friend Peter Morrison. I continued with the same project at Madison. My research expenses were paid by a grant from the Office of Naval Research to Morrison. I drove a gray van that had written on each side, in white paint, U.S. NAVY, which must have seemed strange to see in the middle of the Great Basin desert. My tolerance for the desert was sustained by my occasional retreat to Henderson, Nevada, thanks to the kindness of Fred Ryser, a former student of Peter, and his family. Additional populations of *Peromyscus* were donated by Bill Reeder, H. J. Egoscue, and John Eisenberg, which were greatly appreciated and contributed significantly to this project.

···

corrected for body mass, correlated with climate (McNab & Morrison 1963): individuals from the mesic environment of Berkeley had the highest mass-independent BMRs, those from desert mountain environments had intermediate BMRs, and those from desert lowlands had the lowest BMRs (fig. 7.1). The thermal conductance of these populations also reflected climate: it decreased (thermal insulation increased) in desert populations (fig. 7.2), presumably in compensation for, or at least reflecting, the decrease in BMR. The sum of the change in BMR and conductance in these populations correlated positively with mean ambient temperature and inversely with rainfall (fig. 7.3); that is, with measures that reflect the climate in which these populations lived.

Peromyscus californicus had a much lower mass-independent basal rate than the other species and was restricted to the narrowest range of environmental conditions. This species is large for this genus (50 g) and distinct in that it builds elaborate nests similar to those of the cricetid wood rats (*Neotoma*) (McCabe & Blanchard 1950). Furthermore, *P. californicus* has a mean litter size of 1.9, whereas the other two species collected in Berkeley have larger litters, 5.0 for *P. maniculatus* and 3.4 for *P. truei* (McCabe & Blanchard 1950). The small litter size of *P. californicus* reflects its larger mass and lower mass-independent basal rate (see chap. 14). Indeed, the adjustments made by *P. californicus* almost appear to reflect the response of a desert species that

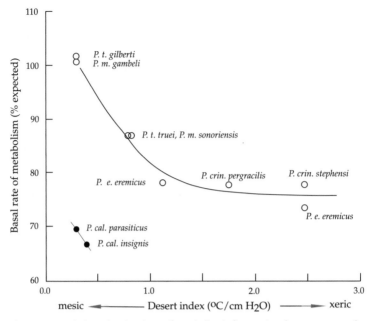

Figure 7.1. Mass-independent basal rate of metabolism in five species of *Peromyscus* as a function of a desert index. Species abbreviations: *P. m.*, *P. maniculatus*; *P. t.*, *P. truei*; *P. cal.*, *P. californicus*; *P. crin.*, *P. crinitus*; *P. e.*, *P. eremicus*. (Modified from McNab & Morrison 1963.)

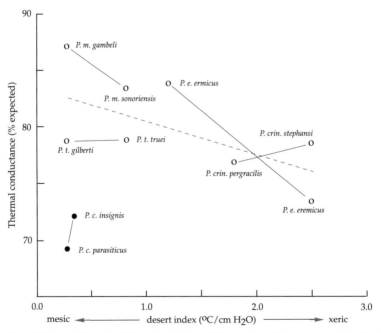

Figure 7.2. Mass-independent thermal conductance in five species of *Peromyscus* as a function of a desert index. Species abbreviations are as in figure 7.1. (Modified from McNab & Morrison 1963.)

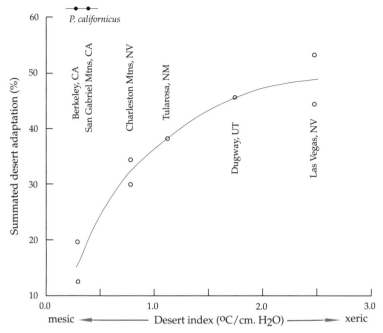

Figure 7.3. Summed desert adaptation, which is defined as the sum of mass-independent basal rate and mass-independent thermal conductance, in *Peromyscus* as a function of a desert index. (Modified from McNab & Morrison 1963.)

has moved into a mesic environment (see fig. 7.3). This species, like the two desert species studied, is a member of the subgenus *Haplomylomys*, so these responses may be reflective of phylogeny and environmental history.

Nearly forty years later, Mueller and Diamond (2001) confirmed the findings of this study in five species of *Peromyscus*, three of which were the same. They showed that, as was implied by McNab and Morrison (1963), variation in the basal rate of metabolism in *Peromyscus*, independent of body mass, is proportional to net community production (fig. 7.4). As in the earlier study, Mueller and Diamond showed that *P. californicus* had an unexpectedly low basal rate (see fig. 7.4), which appears to reflect factors other than body mass, climate, and community production. Mass-independent basal rates of populations of the rodent *Ocotodon degus* in Chile also correlated with habitat productivity (Bozinovic et al. 2009).

The reduction of BMR in desert rodents occurs worldwide, especially in species that feed on seeds (Shkolnik & Borut 1969; MacMillen & Lee 1970; Bowers 1971; Chaffee & Roberts 1971; MacMillen et al. 1972; Scheck 1982; Hinds & MacMillen 1985; Haim & Borut 1986; Haim 1987a,b; Haim & Izaki 1993, 1995; Lovegrove 2000), although a few exceptions exist (Hooper & Hilali 1972), often in species that have a higher insect intake, such as the

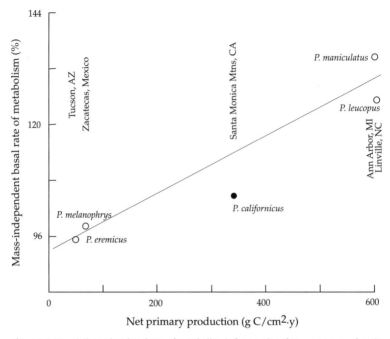

Figure 7.4. Mass-independent basal rate of metabolism in five species of *Peromyscus* as a function of community primary production. (Modified from Mueller & Diamond 2001.)

jerboas *Jaculus deserti* and *J. orientalis*. Invertebrate-eating desert hedgehogs (*Paraechinus*), however, have lower basal rates than hedgehogs (*Erinaceus*) living in mesic environments (Shkolnik and Schmidt-Nielsen 1976), the lower basal rates in hedgehogs possibly reflecting a larger mass than found in jerboas.

The rodent family Heteromyidae has been intensively studied. It is principally found in xeric and desert environments in North and Central America. Heteromyids include pocket mice (*Perognathus, Chaetodipus*), spiny pocket mice (*Liomys*), kangaroo mice (*Microdipodops*), and kangaroo rats (*Dipodomys*), so called because the pocket mice have cheek pockets to store seeds and the kangaroo mice and rats have enlarged hind limbs that permit them to escape perceived threats by hopping when startled.

Heteromyids are generally characterized by low basal rates of metabolism, especially in small species belonging to the Perognathinae (mean BMR = 81% [range 56–109, N = 8] of the values expected from body mass in mammals) (fig. 7.5), which reflects their xeric distributions, their consumption of dry seeds, and their propensity to enter torpor (see chap. 8). All heteromyids that enter torpor have basal rates below the boundary curve, as do four species of kangaroo rats, only one or two of which are known to enter torpor. The two species of *Liomys* fall below the curve, but are not known to enter torpor.

Unlike other heteromyids, species of the genus *Heteromys* are found in tropical rainforests in Central America and northern South America. Two *Heteromys anomalus* were captured at 1,000 m in the cloud forest at Rancho Grande, in coastal Venezuela. This species had a basal rate that was 136% of that expected from body mass (McNab 1979a), which is obviously much higher than the basal rates of desert heteromyids, most of which have basal rates that are about two-thirds of that value. Hinds and MacMillen (1985) demonstrated that another species belonging to this genus (*H. desmarestianus*), which lives in wet forests at 1,400 m in Monteverde, Costa Rica, had a basal rate equal to 121%. Therefore, BMR within this family varies with climate, and possibly with a change in diets from seeds in desert species to a mixed diet of seeds, berries, nuts, and fungi in rainforest species. Ecological factors have obviously had a big impact on energetics in this clade: ultimately, the factors most responsible for the low basal rates of most heteromyids, given their granivory, is the low net primary production of xeric

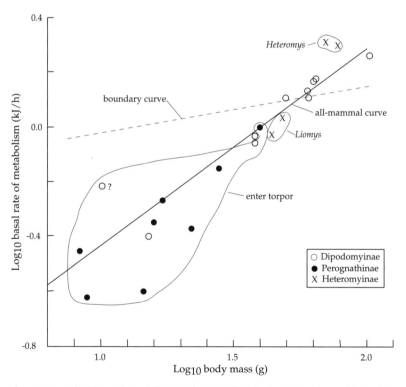

Figure 7.5. Log_{10} basal rate of metabolism as a function of log_{10} body mass in heteromyid rodents in relation to the occurrence of torpor. The boundary curve is indicated. (Data from McNab 2008a.)

and desert communities, which reflects high ambient temperatures and dry conditions during the growing season.

The effects of various factors that influence BMR are often difficult to separate, and that is especially true of climate and food habits. As we just saw in the case of heteromyids, a difference in climate was associated with a difference in food habits. Is the high BMR in *Heteromys* related to climate, food, or both? We have also seen this difficulty in analyzing data from aquatic mammals (see chap. 6): do the high basal rates of marine carnivores reflect life in cold water or carnivory? This difficulty stems from the problem of correlation: the observation that some state or behavior is correlated with a factor does not mean that the state or behavior is directly influenced by that factor. It may simply mean that the state or behavior is associated with another factor that is correlated with the first factor. In other words, we may have entered a causal labyrinth. The answer to these dichotomies may not be *either/or*, but *and*.

This conclusion is well illustrated by an examination of the impact of a desert existence on canids (fig. 7.6). The fennec (*Vulpes zerda*), which weighs only 1 kg and relies heavily on an insectivorous diet, has a low mass-independent basal rate (73%). Williams et al. (2004) concluded that there was no evidence of a climatic effect on the BMR of canids based on data from Arctic, temperate, and desert species and populations. However, Careau et al. (2007) later concluded that there was evidence that desert populations

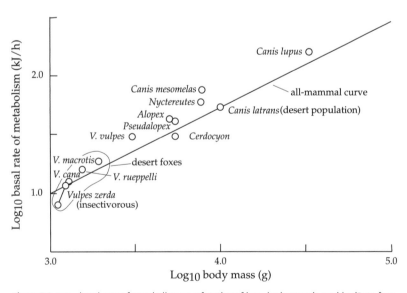

Figure 7.6. Log$_{10}$ basal rate of metabolism as a function of log$_{10}$ body mass in canids. (Data from McNab 2008a.)

of canids had lower BMRs than species and populations living in mesic environments.

How could such a disagreement occur when most of the data used in the two studies were the same? In part, this difference reflected the choice of measurements used in the study: a higher measurement (101%) for the fennec by Maloiy et al. (1982) was used by Williams and colleagues, whereas a lower measurement for this species (73%) by Noll-Banholzer (1979), as well as the higher measurement, was used by Careau and colleagues. This conflict raises a serious question about the use of conflicting data in a comparative analysis because the selection of data may influence the conclusions (see box 7.2).

Another complication in the analysis of character states is the choice of the standard against which to make comparisons. Careau and colleagues fitted a curve of $\log_{10}$ BMR on $\log_{10}$ mass in an attempt to determine the influence of factors such as climate and food habits. However, a difficulty with this approach, as justifiable as it is, is that the fitted curve incorporates all factors inherent in the data, not mass alone. This complication increases as the range in mass decreases. As stated, when there is no variation in mass, all the variation is due to other factors. In the case of the canids, body mass varied by a factor of only $31.6:1 = 10^{1.50}$.

If large species have one habit and small species another, then it is impossible to separate the influences of size and habits, except possibly to say that size and habits are statistically linked, whether they are functionally linked or not. Most of the canids that live in temperate and Arctic environments have intermediate to large masses and feed principally on vertebrates, whereas

BOX 7.2. ·

Choice of Competing Measurements

When I have a choice between two or more conflicting estimates of BMR, my tendency is to use the lower estimate. The usual problem in measuring BMR is that activity or anxiety is more likely to raise the estimate of BMR than is a failure of thermoregulation to reduce it, especially in species that weigh more than 100 g. A failure to regulate body temperature can be excluded by measuring body temperature. I once reported a basal rate for the 900 g aquatic marsupial *Chironectes minimus* that was 121% of the expected value (McNab 1978b), but later Thompson (1988) reported a value for this species equal to 77%, which is the value that I now use, and it is more in accord with other measurements on marsupials. I have no idea why I had a higher value. However, this diversity of measurements raises a concern that must be addressed: as seen in *Peromyscus*, one cannot assume that all populations of a species will have the same BMR (although it is doubtful that the difference in *Chironectes* was populational).

· ·

most of the small-bodied species live in hot deserts and feed on insects (see fig. 7.6). Therefore, the fitted curve contains an environmental and dietary signal as well as the influence of mass. However, if equation (1.4) is used as a mass standard, where mass is the overwhelming factor affecting BMR, given a mass range of $10^{6.17}$, then most of the intermediate- and large-bodied species have high basal rates, including the Arctic fox (*Alopex lagopus*), red fox (*Vulpes vulpes*), wolf (*Canis lupus*), black-backed jackal (*C. mesomelas*), and raccoon dog (*Nyctereutes procyonoides*) (see fig. 7.6). The only exceptions are *desert* coyotes.

Small desert foxes have standard to low basal rates. The difference in interpretation resulting from changing standards is not great, but the wider mass standard undoubtedly gives a clearer indication of the effect of body mass with a reduced influence of food habits and climate. The impact of a desert existence appears to be greatest in small species because the small mass of desert foxes itself reduces energy expenditure, which permits an increased reliance on insects as the principal food item in a resource-limited environment.

A decrease of BMR in mammals living in xeric environments reduces their evaporate water loss by reducing gas exchange in a dry atmosphere, an appropriate response for balancing a water budget in a desert. Of course, one of the most distinctive responses of mammals to life in a desert environment is the evolution of the capacity to produce highly concentrated urine (McNab 2002a). This capacity is especially noteworthy in seed-eating rodents, which have a low water intake. Whether the capacity to produce high urine to plasma ratios is linked in some manner to energy expenditure has not been examined, although some of the desert rodents most noted for this capacity are members of the family Heteromyidae, which also show the greatest reduction in BMR. No association of energy expenditure with the production of concentrated urine appears to exist, unless the low basal rates of heteromyids might have been even lower if they did not have the capacity to produce highly concentrated urine.

Birds

For a long time, it appeared that birds were able to survive life in dry, even desert, environments without making substantial physiological adjustments in either energy expenditure or water conservation. Bartholomew and Cade (1963) suggested that the principal ways in which birds tolerated a desert existence were by (1) remaining close to a water source, (2) having extended mobility so that they could return to a water source, and (3) in the case of insectivorous or carnivorous species, depending on the water available in their food. But these authors also noted that the New World deserts they studied, especially in California, were not ancient; a greater adjustment therefore might have been made by birds to life in Old World deserts, which have a longer history. (How long does adaptation take? Does the environment change at a sufficiently slow rate that some species can successfully track the changes?)

Early work on the question of adaptation of birds to life in Australian deserts also showed no remarkable physiological adjustments. Although these studies were principally focused on water balance, no remarkable differences in energy expenditure were found between xerically and mesically distributed birds. Furthermore, an extended study of the energetics of eleven species of Australian parrots (Williams et al. 1991) and a comparison of four species of African lovebirds (*Agapornis*) with four species of Australian grass parrots (*Neopsephotus* and *Neophema*) (Burton et al. 2008) found no differences in BMR between mesically and xerically distributed species. Maybe the high flight velocities of parrots permit them to use distant localized water sources.

Ever so gradually, some birds that live in hot, dry environments have been shown to have lower basal rates of metabolism than similar species that live in mesic environments. For example, horned larks (*Eremophila alpestris*) from North America (Trost 1972) and spinifex pigeons (*Geophaps plumifea*) from Australia (Dawson & Bennett 1973; Withers & Williams 1990), both species living in dry grasslands or rocky arid regions, have lower basal rates than expected from body mass (75% and 68%, respectively). The only difficulty with these studies is that they were of single species and therefore lacked an evolutionary context.

More work on selected groups of birds has recently demonstrated their reduction of energy expenditure in xeric environments (Chaffee & Roberts 1971). Much of this work was accomplished through the persistence of Irene Tieleman and Joseph Williams. An extended analysis of the energetics of twelve species of larks (Alaudidae) that were found along an aridity index in the Arabian desert showed that their mass-independent BMRs decreased with an increase in aridity (fig. 7.7); this work was in some ways a great extension of the work on larks started by Trost (1972). (The aridity index used, which is a function of the annual precipitation, the mean maximal temperature of the hottest month, and the mean minimal temperature of the coldest month, is lowest in the driest environments and is therefore opposite to the aridity index used in figs. 7.1, 7.2, and 7.3.) Body mass accounted for only 53% of the variation in BMR; this small fraction resulted from a small range in body mass (from only 15.2 to 50.6 g, a 3.3:1 range) and a large influence of environmental conditions. When aridity was brought into the analysis with body mass, 91% of the variation in BMR was accounted for (Tieleman et al. 2002; Williams & Tieleman 2005). A satisfying aspect of this

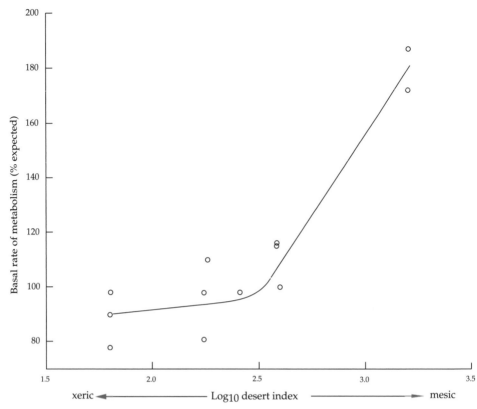

Figure 7.7. Basal rate of metabolism in larks (Alaudidae) as a function of a desert index. (Modified from Tieleman et al. 2002.)

correlation is that it occurs along an aridity gradient, rather than the environmental conditions being artificially broken into two arbitrary categories, mesic and xeric.

Seasonality in dry environments is an important factor influencing water availability and energy expenditure. The white-browed scrub-wren (*Sericornis frontalis*), a 11.0 g Australian passerine, has a lower mass-independent basal rate in xeric (86%–95%) than in semiarid (106%–124%) or mesic environments (101%) in summer, whereas in winter no such differences (109%–116%) were found (Ambrose & Bradshaw 1988). These variations in BMR reflect seasonal and environmental differences in rainfall and insect abundance. Two species of this genus that are endemic to the wet, cool tropical highlands of New Guinea have basal rates that are similar to that of the Australian species in semiarid or winter conditions, 100% (*S. perspicillatus*, 8.5 g) and 105% (*S. nouhusyi*, 16.1 g), although another species, *S. papuensis* (9.8 g), has a lower BMR, 92% (McNab 2009a).

A factor influencing BMR is the variability of conditions in the environment. Cavieres and Sabat (2008) examined the response of BMR in rufous-crowned sparrows (*Zonotrichia capensis*) to the absolute value of aridity and its variation in Chile. They demonstrated that the response of BMR in this species to thermal acclimation in the laboratory varied with the aridity of the environment from which the birds came. For example, the BMRs of individuals from different populations varied little when they were acclimated to 30°C. At 15°C, individuals from the arid, stable environment at Copiapo showed little change in BMR, but BMR increased greatly in individuals from the climatically arid and variable environment of Santiago, and increased even more in individuals from the highly variable but cool, wet environment of Valdivia (fig. 7.8). In this case, thermal acclimation was most prevalent in the population that faced the greatest variation in rainfall, rather than in mean temperature.

A general analysis by Tieleman and Williams (2000) indicated that the

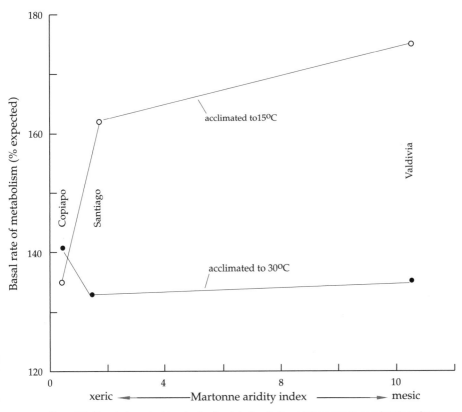

Figure 7.8. Mass-independent basal rate of metabolism in *Zonotrichia capensis* as a function of an aridity index when acclimated to two temperatures. (Modified from Cavieres & Sabat 2008.)

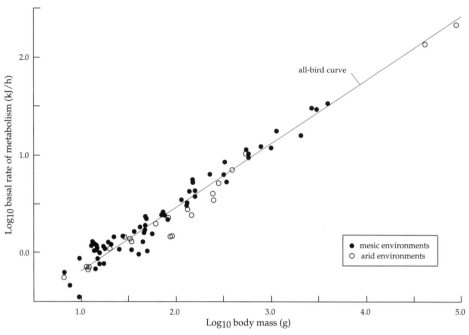

Figure 7.9. Log$_{10}$ basal rate of metabolism as a function of log$_{10}$ body mass in birds living in mesic and arid environments. The all-bird curve is indicated. (Modified from Tieleman & Williams 2000.)

basal rates of birds living in deserts average 83% of the values found in non-desert species (fig. 7.9), although a potentially better comparison would be with terrestrial flighted species because the study's inclusion of procellarii-forms and penguins adds confusing factors, as does the inclusion of large flightless birds, which have inherently low basal rates (see chaps. 3 and 9). The analysis also showed that field rates of metabolism in desert species average 51% of the mean in non-desert species (see chap. 11).

Life in a warm, moist environment

Repeated measurements have shown that tropical birds generally, and tropical passerines in particular, have lower basal rates than their temperate equivalents or relatives (Wikelski et al. 2003a; Wiersma et al. 2007; McNab 2009a). Many tropical mammals also have low basal rates, including marsu-pials, tree sloths, anteaters, armadillos, and some primates (McNab 2008a). Some tropical mammals with low basal rates are referred to as being "slow," which implies not only that they are comparatively inactive, or even seden-tary, but also that they have low rates of reproduction, slow rates of growth, long life spans, and often are solitary.

The usual explanation for the low basal rates of tropical animals has been that they reflect life in a warm, moist environment: tropical species do not have to elevate their basal rates to compensate for low ambient temperatures, as might be required in a cool to cold environment. That is, their low BMRs were interpreted as a thermoregulatory response to life in a tropical climate. Furthermore, some lifestyles and food habits, as well as phylogenetic characteristics (see chap. 13), in warm, humid environments are associated with a reduction in basal rate. Some of these factors are arboreal, sedentary, and fossorial lifestyles.

An arboreal lifestyle

Low basal rates are especially common among tropical arboreal mammals, although their pattern of occurrence reflects the animals' approach to life. Low basal rates may not directly reflect life in trees versus a terrestrial lifestyle, but are probably based on activity level, which may itself reflect differences in body composition and food habits. Sluggish behavior in a terrestrial mammal, however, would probably increase its vulnerability to predators unless it had some passive armor, as is found in echidnas, hedgehogs, armadillos, and pangolins. Active arboreal mammals include rodents and larger primates, whereas sluggish species include marsupials, sloths, some anteaters, some smaller primates, a few rodents, and most carnivorans.

Active arboreal species that have been measured generally have high BMRs (fig. 7.10). Active arboreal primates, including *Saimiri* and *Cercopithecus*, have BMRs between 132% and 143% of the value expected from mass. Among the rodents measured are various tree squirrels, including *Tamiasciurus* and *Sciurus*, which have BMRs equal to 119% and 141%, and the arboreal arvicoline rodent *Arborimus pumo*, which weighs only 20 g and has a BMR of 183%. These species are active and small, and both factors permit a high mass-independent basal rate. Some larger primates (*Pan troglodytes*, 141%, *Macaca mulata*, 115%) spend time both in trees and on the ground. The same is probably true of meat-eating carnivores that spend some time in trees, including the clouded leopard (*Neofelis nebulosa*), leopard (*Panthera pardus*), and jaguar (*P. onca*). The jaguar is the only species in this group that has been measured; it has a BMR equal to 131% of the value expected from mass, which is similar to the values seen in strictly terrestrial felids (McNab 2000c). A similar association between a mixed arboreal/terrestrial lifestyle and high basal rates is found in the Eurasian marten (*Martes martes*), which has a BMR of 154%, and the North American marten (*M. americana*), with a BMR of 131%. However, some tropical squirrels that spend time both in trees and on the ground (*Funisciurus, Paraxerus, Heliosciurus*) have intermediate BMRs, generally between 76% and 107%.

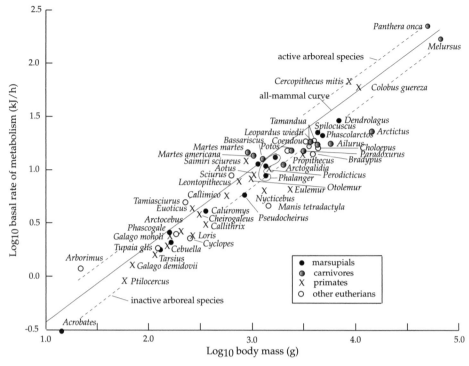

Figure 7.10. Log_{10} basal rate of metabolism as a function of log_{10} body mass in active and inactive arboreal mammals. The all-mammal curve and active and inactive arboreal curves are indicated. (Data from McNab 2008a.)

Inactive, or sedentary, arboreal mammals, in contrast, have much lower rates of metabolism. For example, arboreal marsupials, especially cuscuses (*Phalanger*, *Spilocuscus*) and brush-tailed possums (*Trichosurus*), which feed on a mixture of fruit and leaves, have basal rates that vary between 68% and 90% (McNab 2008a) (see fig. 7.10). The sedentary, arboreal, folivorous koala (*Phascolarctos cinereus*) has a low BMR, at 67%, as does the ring-tailed possum (*Pseudocheirus peregrinus*), at 62%. An arboreal tree kangaroo (*Dendrolagus matschiei*), which feeds on fruit and leaves, has a BMR of 71%, which is appreciably lower than those of terrestrial kangaroos (83%–96%).

Several species among the Neotropical Pilosa are committed to a sedentary, arboreal existence (see fig. 7.10). These species include the folivorous tree sloths (*Bradypus*, *Choloepus*) and arboreal ant and termite eaters (*Cyclopes*, *Tamandua*). Tree sloths, of course, are the epitome of a sedentary, "slothful," existence. So, it is not surprising that they have BMRs that vary between 52% and 55% of the values expected from mass—so low, in fact, that they require a thick fur coat (minimal thermal conductance being 56% to 65% of the values expected from mass [McNab 1978c]) to compensate for their low heat production, but still have a compromised capacity for temperature regulation, which is partly mitigated by their having masses of 3.5 to 4.5 kg (McNab 1978c). *Cyclopes didactylus,* an arboreal anteater, has a BMR of 62% (McNab 1984), which, given its mass of only 250 g, compromises its capacity for temperature regulation at ambient temperatures <15°C. Two other species of sluggish arboreal anteaters, *Tamandua mexicana* and *T. tetradactyla,* also have low basal rates (73% and 70%, respectively) (Enger 1957; McNab 1984).

Inactive arboreal primates include the Philippine tarsier (*Tarsius syrichta,* which has a BMR equal to 73%); tupaids (*Tupaia,* 81%–84%); a ptilocercid (*Ptilocercus lowii,* 67%); lorisids (*Arctocebus, Loris, Nycticebus, Perodicticus,* 51% to 81%); a lemur (*Eulemur fulvus,* 35%); and an indriid (*Propithecus verreauxi,* 60%) (Bradley & Hudson 1974; Whittow & Gould 1976; Palacio 1977; Müller 1979; Weigold 1979; Daniels 1984; Müller et al. 1985; McNab & Wright 1987; Richard & Nicoll 1987). Arboreal primates belonging to the family Galagidae, including *Galago, Euoticus,* and *Otolemur,* have intermediate levels of activity and BMRs, 84% to 111% of the values expected from mass (Hildwein 1972; Müller & Jaksche 1980; Döbler 1982). The somewhat sedentary *Colobus guereza,* which is principally a folivore, weighs 10.5 kg and has a BMR of 108%. A comparison of two Neotropical primates would probably confirm this dichotomy in energetics: the active spider monkey (*Ateles* spp.), which feeds on fruit and leaves, undoubtedly has a higher BMR than the howler monkey (*Alouatta* spp.), an inactive arboreal folivore. Indirect evidence of this proposed difference is the appreciably thicker fur coat of the howler, even though both monkeys share the same rainforest environments.

Some members of the order Carnivora are committed to an arboreal lifestyle, many of which feed on a mixture of small vertebrates, fruit, and large insects, or in some cases leaves (see chap. 5). These animals include *Nandinia* (Nandinidae) from Africa; *Arctogalidia, Paradoxurus,* and *Arctictis* (Viverridae) from southern Asia; and *Bassariscus* and *Potos* (Procyonidae) from Central and South America. Most of these carnivorans have BMRs that vary between 60% and 80% (Müller & Kulzer 1977; McNab 1995) (see fig. 7.10), although body mass and their responses to food habits and arboreality are interconnected. The small, omnivorous *Bassariscus sumichrasti,* at 1.3 kg, has a BMR of 105% (Chevalier 1987). In contrast, the large *Arctictis binturong,* at 14 kg, which feeds principally on fruit and leaves, has a BMR of only 33% (McNab 1995). The red panda (*Ailurus fulgens*), from Asia, feeds principally on bamboo shoots and leaves; at 5.7 kg, it has a BMR that is only 49% of the value expected from mass (McNab 1988a).

The BMRs of arboreal species are probably confounded by their membership in various phyletic groups, their sedentary habits, and their consumption of vertebrates, fruit and leaves, or ants and termites. However, when the $\log_{10}$ BMR of arboreal species is plotted as a function of $\log_{10}$ mass (see fig. 7.10), it is remarkable that all inactive species, including marsupials, fall along the same curve, which is about 66% of the values expected from the all-mammal curve, with much variation; the most shared factor is a sedentary lifestyle, possibly in association with food habits. These data reinforce the conclusion (McNab 1978b, 2005b) that when external factors depress energy expenditure, the dichotomy between marsupials and eutherians is diminished, or even eliminated (see chap. 13), just as it was among ant and termite eaters. As we have seen, active arboreal mammals have basal rates that are *higher* than expected from the all-mammal curve, but that category never includes marsupials (see fig. 7.10). However, this division of species into active and inactive is a simplification, for there is probably a graded range of activity levels and basal rates among arboreal mammals.

Especially interesting here is the observation that the one measured arboreal carnivore committed to eating vertebrates, the margay (*Leopardus wiedii*), has a BMR that is only 73% of the value expected from mass (McNab 2000c), which conforms to the BMR of the other arboreal species (see fig. 7.10). The margay's close relative, the ocelot (*L. pardalis*), a terrestrial felid, has a BMR equal to 114%, 56% greater than that of the margay (see chap. 5). The difference in energy expenditure between these two species correlates with a difference in morphology: the ocelot is a stout, highly muscular species, whereas the margay is a sleek, thin cat. This comparison suggests that an important adjustment to an arboreal lifestyle, independent of food habits and phylogeny, is a reduction in BMR associated with a reduction in activity and in heart and muscle masses.

Tropical birds

The equality of basal rates in four avian orders with the basal rate of the passerine collective (see chap. 3) led to the suggestion (McNab 2009a) that tropical passerines may have lower BMRs than temperate passerines because tropical species are generally sedentary, not because they live in warm environments, although sedentariness itself is not independent of climate: no polar birds are sedentary, including gyrfalcons (*Falco rusticolus*), snowy owls (*Nyctea scandiaca*), and redpolls (*Carduelis* spp.).

As interesting as this migratory-sedentary dichotomy might be, the observation that among frugivorous specialists, most of which are tropical residents and presumptively nonmigratory, passerines still have higher mass-independent basal rates (see fig. 5.9) suggests that a passerine difference from other birds remains, even when only tropical species are considered. A further question remains unanswered: do altitudinal migrants in the tropics have higher basal rates than either lowland or highland endemics?

A further complication is that the distributions of many avian orders and passerine families are limited to the moist tropics, and few species that belong to these groups have had their energy expenditures measured. Some of these groups in the Neotropics include Tinamiformes, Cathatidae, Cracidae, Aramidae, Psophiidae, Cariamidae, Psittaciformes, Cuculiformes, Trochilidae, Trogoniformes, Coraciiformes, Piciformes (except for Ramphastidae), and among the Passeriformes, Dendrocolaptidae, Furnariidae, Formicariidae, Tyrannidae, Pipridae, Cotingidae, Muscicapidae, Emberizidae, and Vireonidae. A similar list could be established for all other tropical regions. Knowledge of the energetics of many species that belong to these groups will undoubtedly improve our understanding of the energetics of tropical birds and the rules that apply to birds in general. For example, recent work on the energetics of kingfishers from New Guinea indicated a propensity of some species to permit their body temperatures to fall to unusually low levels in association with low mass-independent basal rates (McNab, personal observations). This may simply be the tip of a tropical iceberg.

A fossorial existence

Many fossorial mammals encounter a warm, moist environment. Only a minority of terrestrial mammals have evolved a fossorial existence, although the fossorial lifestyle is phyletically and geographically widespread. These animals include a marsupial "mole" (*Notoryctes typhlops*) from Australia; golden moles (Chrysochloridae) from Africa; moles, shrew-moles, and desmans (Talpidae) in the Holarctic; and a series of rodents. Fossorial rodents include the mole-lemming (*Ellobius*) and zokor (*Myospalax*) in the Cricetidae from Asia; blind mole-rats (*Spalax*, Spalacidae) from the Middle East; bamboo rats (*Rhyzomys*, Spalacidae) from Asia; African mole-rats (*Tachyoryctes*, Spalacidae, and *Heterocephalus*, *Heliophobius*, *Cryptomys*, Bathyergidae) from Africa; pocket gophers (Geomyidae) from North and Central America; the cururo (*Spalacopus cyanus*, Octodontidae) from Chile; and tuco-tucos (*Ctenomys*, Ctenomyidae) from southern South America. The mountain "beaver" (*Aplodontia rufa*) from northwestern North American may be (marginally) included. The extent to which two "fairy" armadillos (pichiciegos), *Chlamyphorus truncatus* from central Argentina and *Calyptophractus retusus* from Bolivia and Paraguay, are fossorial is uncertain, although all armadillos are burrowers, even the giant armadillo (*Priodontes maximus*), which weighs up to 60 kg.

Many adjustments are required of mammals that make a commitment

to a fossorial life (Nevo 1999). It is a commitment to life in a closed burrow system, a hypoxic, hypercapnic, still- gaseous environment with high
relative humidities and intermediate to warm temperatures (Arieli 1990).
The energetics of fossorial mammals has been the subject of extensive attention (McNab 1966; Gorecki & Christov 1969; Klein 1972; Bradley et al.
1974; Bradley & Yousef 1975; Gettinger 1975; McNab 1979b; Ross 1980;
Contreras 1986; Lovegrove 1986a,b, 1987; Contreras & McNab 1990). Most
fossorial rodents respond to burrow conditions with a reduction in BMR
and an increase in thermal conductance. The reduction of BMR in fossorial
mammals varies with body size and the conditions encountered in burrows,
the atmospheres of which are generally warmer and more humid than the
external environment, principally because all fossorial mammals plug their
burrows to reduce predation. Reductions of BMR are seen mainly in species
that weigh more than 60 – 80 g (Contreras & McNab 1990), which means
that fossorial insectivores, most of which are smaller than 70 g, tend to have
BMRs that are greater than expected from body mass (fig. 7.11), with the
exception of the marsupial *Notoryctes typhlops*.

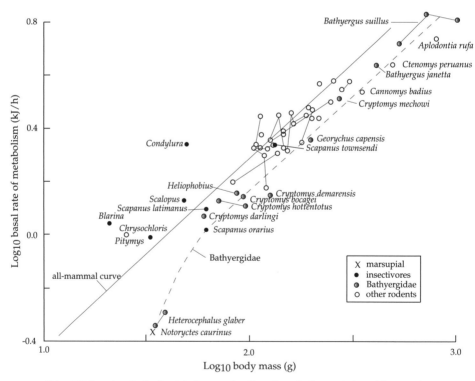

Figure 7.11. Log_{10} basal rate of metabolism as a function of log_{10} body mass in fossorial mammals.
(Data from McNab 2008a.)

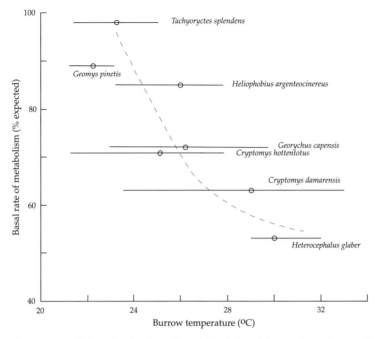

Figure 7.12. Mass-independent basal rate of metabolism in fossorial mammals as a function of burrow temperature. Range in burrow temperatures is indicated by the horizontal line. (Data from McNab 1979b.)

The reduction of basal rate in fossorial rodents is carried to the extreme in the naked mole-rat (*Heterocephalus glaber*), which often faces burrow temperatures greater than 30°C (fig. 7.12). The claim that fossorial mammals usually have high thermal conductances (McNab 1966, 1979b) is only marginally correct, but is most clearly the case in very warm burrows, which reach their extreme in the case of *Heterocephalus* (fig. 7.13). Its relative, *Cryptomys damarensis*, is not confined to high temperatures in summer burrows, given the great range of temperatures within its burrow system. *Heterocephalus*, however, cannot escape high burrow temperatures, which sets the stage for its radical response to a fossorial existence in conjunction with a high-density social system (see box 7.3).

In Ethiopia, Somalia, and northern Kenya, the naked mole-rat faces high burrow temperatures (29°C–32°C) and relative humidities (92%–100%) (McNab 1966). Under such conditions, an endotherm potentially faces death. *Heterocephalus* has responded to this threat with a reduction in its body mass (39 g) compared with those of other bathyergids (which weigh from 60 g to 1.0 kg); a reduction in the mass-independent rate of metabolism (53% of that expected from mass), and a marked increase in mass-independent conductance (240%) that results from a complete loss of

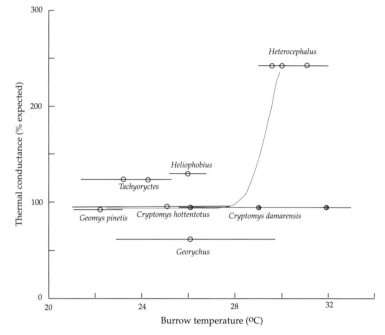

Figure 7.13. Mass-independent thermal conductance in fossorial mammals as a function of burrow temperature. (Data from McNab 1979b.)

BOX 7.3. ··

Collecting Naked Mole-Rats

I went to Kenya in 1965 to capture naked mole-rats, a trip that was partially financed by the American Philosophical Society. I had to give assurance to the Society that (1) I would try to stay alive (I got to Nairobi one month after *Uhuru*—independence—and the national stadium was still bedecked with buntings) and (2) I would stay as long as it would take to get *Heterocephalus*. While in Kenya, I was greatly helped by many people, including J. Leakey, R. Bloom, and R. Percival.

John Seago, who had a game camp in Isiolo, warned a group of us seated around a campfire at night that our obvious presence might be an invitation to the *shifta*, Somali raiders, potentially located on some nearby ridges, to use us for target practice. (A foretelling of the future?)

The ground was exceedingly hard where I encountered *Heterocephalus*. In compensation, I shaved the ground above a burrow, which would permit me to thrust the shovel through the burrow to trap a mole-rat that was investigating an opened plug. When a burrow was opened, naked-mole rats formed a chain gang, passing soil from one to another and finally to the last individual, who plugged the burrow. In addition, a rotation among the individuals forming the chain occurred, so that the diggers and the pluggers were not always the same individuals. The first time I thrust the shovel, I caught 7.5 animals! In 1965, nothing was known about the communal nature of this mole-rat's society (Jarvis 1981; Jarvis et al. 1994).

In Florida, the animals were kept in an aquarium with sand and a heat lamp at one end; they would usually form a pile under the heat lamp. Archie Carr, a mischievous member of my department, famous for his nature writing and his advocacy for the conservation of sea turtles, would place a ping-pong ball in the aquarium with the animals and would then bring people into my lab in excitement, telling them that this "primitive" animal was laying eggs! Some of the unwary went away confused.

. . . .

pelage. These changes have reduced its BMR/C ratio from the standard $1.00/1.00 = 1.00$ to $0.53/2.40 = 0.22$; that is, $\frac{1}{5}$ only of the value expected from mass. As a consequence, the naked mole-rat is an incompetent homeotherm (see fig. 1.9).

The naked mole-rat never encounters low ambient temperatures in its equatorial environment (see chap. 12). In fact, the artificial nature of its low body temperature, which occurs only under cold exposure in the laboratory, is demonstrated by the congealing of its body fat at low body temperatures, which can be reversed only by "melting" when its body temperature is artificially increased. *Heterocephalus* cannot spontaneously arouse from a low body temperature. Given that 29°C is probably the lowest ambient temperature *Heterocephalus* faces in its burrow system, 32°C is probably its lowest body temperature in the field (see fig. 1.9A), at which its rate of metabolism is maximally twice the basal rate (see fig. 1.9B).

CHAPTER EIGHT. *Evasions*

Birds and mammals have been forced by physics, ecology, and behavior to adjust their normally high energy expenditures to conform to the limited energy availabilities imposed by various external and internal factors. Some of these adjustments have been modest, such as being active under leaf litter or snow layers. Others include seasonal variations in rate of metabolism and thermal conductance. Modest adjustments, however, may not suffice in the face of large seasonal changes in environmental conditions and food availability. Some species make more radical, even extreme, adjustments, including seasonal modifications of their geographic distribution to evade harsh conditions. Other species temporarily abandon their commitment to a high body temperature, especially those that do not have the capacity for long-distance movement. The two principal evasions, then, are migration and entrance into torpor, both of which permit species to avoid environmental conditions that exceed their capacity to face with modest adjustments of behavior and energy expenditure.

Migration

A common response to seasonally harsh climates or restricted food supplies is to move to a more favorable environment. This behavior requires an effective means of moving an appropriate distance. The only endotherms that can make such extended movements are birds, a few bats, some marine mammals, especially baleen whales, and some large herding mammals. Small mammals other than bats cannot make such movements, and even most temperate bats are limited to local movements.

Mammals

Although almost all temperate bats are insectivorous, and although most move from summer roosts to local caves for hibernation, a few species migrate to varying extents. These species belong to two groups. One group comprises the limited number of cool- to cold-temperate species that migrate to warm-temperate or tropical environments for winter. In North America, such species include members of the genera *Lasiurus*, especially *L. borealis* and *L. cinereus*, and *Lasionycteris noctivagans*, and in Eurasia, members of the genus *Nyctalus*, including *N. noctula* and *N. leisleri*. These species are in many ways quite different from the typical north-temperate bats, as represented by *Myotis*. *Lasiurus* is characterized by large litter sizes (Banfield 1974) and has a greater commitment to endothermy than most vespertilionids. Unlike most *Myotis*, *Lasiurus* can have a defined basal rate because of its regulated body temperature and well-developed fur coat (Genoud 1993). *Nyctalus* may have some of the same characteristics and is known to feed extensively on migrating birds (Dondini & Vergari 2000; Ibáñez et al. 2001; Popa-Lisseanu et al. 2007).

The second group of migratory bats consists of a few tropical insectivorous species whose distributions marginally extend into warm-temperate environments. In North America, they include members of the Molossidae and Phyllostomidae. For example, populations of the predominantly tropical free-tailed bat *Tadarida brasiliensis* are migratory in Texas, Oklahoma, Colorado, and New Mexico, where they encounter harsh winters. Nonmigratory populations of this species reside in Florida, Arizona, and California, where they face moderate winters with an abundance of flying insects throughout the year (Cockrum 1969). Migration in this species is facilitated by the high aspect ratio of its wings and its high flight velocities. Three other molossids belonging to the genera *Tadarida* and *Eumops* move seasonally into western North America, and *Eumops glaucinus* has established a permanent population in subtropical southern Florida. A few other tropical bats marginally cross the border from Mexico, especially the insectivorous phyllostomid *Macrotus californicus*, most of which return to Mexico for the winter. Flying foxes (Pteropodidae) in Australia move about in relationship to the monsoonal/dry season cycle (Tidemann et al. 1999; Vardon et al. 2001).

Some large mammals migrate to compensate for seasonal variations in food and water availability. These animals include caribou (*Rangifer* spp.) and musk oxen (*Ovibos moschatus*) on the Arctic plain, ungulates in eastern and southern Africa, and formerly bison (*Bison bison*) on the Great Plains of North America. Marine mammals have seasonal migrations that are timed principally in relation to the presence of open water and the availability of food. For example, humpback whales (*Megaptera novaeangliae*) gorge on krill in Arctic waters, but feed little, if at all, on their wintering grounds near Hawaii.

Birds

Many groups of birds are migratory, which permits them to harvest resources in polar, temperate, and montane environments where permanent residence is not possible, given the food habits and behavior of the species. These species, as noted, belong principally to five orders (Anseriformes, Pelecaniformes, Procellariiformes, Charadriiformes, and Passeriformes). The studied species that belong to these orders have basal rates that, when corrected for mass, are indistinguishably high (McNab 2009a). The evidence that may really tie the high BMRs of these orders to their migratory habits is that in at least three of these orders, sedentary species have lower mass-independent basal rates than migratory species. However, a sedentary member of the Procellariiformes, all of which feed at sea, is difficult to imagine.

Sedentary New Zealand ducks (McNab 2003a) (see chap. 9) have lower basal rates than migratory species belonging to the same genera. Endemic anseriforms in Hawaii, including the koloa (*Anas platyrhynchus wyvilliana*), Laysan duck (*A. p. laysanensis*), and nene (*Branta sandvicensis*), are therefore expected to have low BMRs. A comparison of the koloa and Laysan duck with continental mallards (*A. p. platyrhynchus*) would be especially interesting, given their close relationship and the mobility of continental mallards. A similar comparison should be made between the nene and its close relative, the Canada goose (*B. canadensis*).

Among pelecaniforms, the American anhinga (*Anhinga anhinga*) is generally sedentary, whereas most cormorants are mobile. This difference is reflected in their mass-independent basal rates. The anhinga has a basal rate that is 85% of the value expected from mass (Hennemann 1983), whereas the widespread double-crested cormorant (*Phalacrocorax auritis*) has a basal rate of 125% (Hennemann 1983), and the shag (*P. aristotelis*) and northern gannet (*Morus bassanus*) from the North Atlantic have BMRs equal to 172% and 185%, respectively (Bryant & Furness 1995). Do white pelicans (*Pelecanus erythrorhynchus*) have higher basal rates in association with their greater migratory movements than comparatively sedentary brown pelicans (*P. occidentalis*)?

Do the sedentary species that belong to the other two orders also have low BMRs? A series of endemic charadriiforms are found in New Zealand, including the variable oystercatcher (*Haematopus unicolor*), black stilt (*Himantopus novaezelandiae*), banded dotterel (*Charadrius bicinctus*), New Zealand dotterel (*C. obscurus*), shore plover (*Thinornis novaeseelandiae*), New Zealand snipe (*Coenocorypha aucklandica*), and wrybill (*Anarhynchus frontalis*). Given the observations on endemic New Zealand ducks, these shorebirds might also be expected to have lower basal rates than highly migratory species from the Northern Hemisphere, if sedentariness is a factor that depresses avian BMR or if mobility increases BMR. The one sedentary "shorebird" measured is the least seed-snipe (*Thinocorus rumicivorus*), which has a restricted movement in winter from the Andean altiplano to the lowlands of Uruguay and Patagonia. It has a BMR that is only 56% of the value expected from mass (Ehlers & Morton 1982)! However, the characteristics of this species are peculiar among "shorebirds," including living in a desert and feeding on seeds, leaves, and buds. These characteristics probably influence its rate of metabolism. This observation reemphasizes the value of comparing ecologically divergent species that are closely related.

Limited evidence indicates that many of the Northern Hemisphere shorebirds that undertake long-distance migrations have higher mass-independent basal rates than more sedentary shorebirds. The migratory species include the Eurasian golden plover (*Pluvialis apricaria*, 117%), black-bellied plover (*P. squatarola*, 129%), ruddy turnstone (*Arenaria interpres*, 122%) and green sandpiper (*Tringa ochropus*, 122%). More sedentary species include the Eurasian (*Scolopax rusticola*, 103%) and American woodcocks (*S. minor*, 98%). However, some long-distance migrants have low basal rates, such as the little ringed plover (*Charadrius dubius*, 100%) and the red knot (*Calidris canutus*, 91%).

An extensive examination of the energetics of Palearctic shorebirds with respect to migratory distance gives a picture quite different from the one suggested above (fig. 8.1): species that nest in the Arctic and winter in Africa have basal rates that are ≥20% lower than those of species that winter in Europe. This difference is found even among populations within some species, such as the knot (*Calidris canutus*). No difference was found in basal rate between tropical endemic shorebirds and those Palearctic species that winter along the tropical African coast (Klaassen et al. 1990; Kersten et al. 1998). Palearctic shorebirds that winter in South Africa, however, tend to have basal rates similar to those of species that winter in Europe. These data suggest that the principal factor dictating BMR in migratory shorebirds may be the physical conditions encountered in winter.

The basal rates of Arctic shorebirds are highest during the summer peak in body mass, reflecting a hypertrophy of muscles and digestive organs (Piersma et al. 1995, 1996), not necessarily an increase in mass-independent BMR. Hypertrophy of the pectoral muscles occurs during premigratory fat deposition (Piersma et al. 1996). Shorebirds use these muscles as an energy source for migratory flight (Lindström et al. 2000). Fundamental changes in basal rate reflect different compromises among competing demands at different times of the year (Klaassen et al. 1990), and "variation in (functional components of) lean mass is the vehicle for seasonal adjustments in metabolic physiology to variable demand levels" (Piersma et al. 1996).

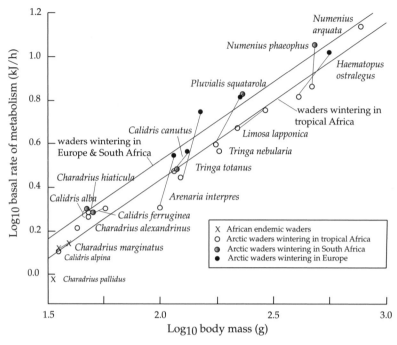

Figure 8.1. Log_{10} basal rate of metabolism as a function of log_{10} body mass in Arctic shorebirds wintering in Europe and Africa and shorebirds endemic to Africa. (Modified from Kersten et al. 1998.)

Obviously, we need many more data on shorebirds, especially on sedentary species. For example, some populations of Wilson's plover (*Charadrius wilsonia*) are quite stationary in the Gulf states, as are some populations of the snowy plover (*C. alexandrinas*). Other populations of the snowy plover nest in the Midwest and winter in the Gulf, as the piping plover (*C. melodus*) does, and thus have an intermediate migratory distance.

The remaining avian order with high basal rates is the Passeriformes. Temperate species have higher basal rates on average than the tropical species studied (Wikelski et al. 2003a; Wiersma et al. 2007; McNab 2009a), although many passerines live in a temperate or polar environment only during spring and summer, when they are not exposed to low ambient temperatures. Their high basal rates may reflect the cost of seasonal migration. In theory, this hypothesis can be examined in several ways. For example, do sedentary temperate passerines have higher basal rates than sedentary tropical passerines? Ideally, this question should be examined in species that belong to the same genus, or at least to the same family, which therefore limits the analysis to three families: Troglodytidae, Meliphagidae, and Fringillidae. This restriction would minimize other factors that may obscure a definitive analysis. Unfortunately, few tropical passerines that belong to these families have been measured. Indeed, the absence of data on the physiology of tropical birds is principally responsible for our inability to perform a general analysis of the factors that determine, or at least influence, avian energetics. The generally low basal rates of tropical passerines may in fact reflect a sedentary way of life rather than as an adaptation to a warm tropical environment (McNab 2009a).

Another approach would be to compare temperate passerines that migrate with temperate passerines that do not migrate. With the presently available data, this question can be marginally examined in the New World in only one family, the Fringillidae. In that case, five migratory species had a mean mass-independent basal rate equal to 136% ± 4%, whereas three nonmigratory species had a mean of 130% ± 12%; in other words, there was no difference. We clearly need many more data from a variety of species and families before these questions can be answered.

The possibility that a low BMR in tropical passerines is related to sedentary behavior was examined in the stonechat (*Saxicola torquata*) by Wikelski and associates (2003a). They examined four populations of this species, a sedentary population from Kenya, a partially sedentary population from Ireland, and migratory populations from Austria and Kazakhstan. All individuals were hand raised to avoid a direct climatic influence on the birds. Sedentary Kenyan birds had the lowest resting rate of metabolism, followed by Irish birds, with the highest rates in the migratory birds from Austria and Kazakhstan. These differences were in birds that were not exposed to conditions in their native environments, which suggests that these differences were likely to be genetically based, and therefore that the low BMRs found in many tropical species reflect a sedentary lifestyle. A similar result in this species was found by Klaassen (1995).

The data on ducks, passerines, and possibly the pelecaniforms raise a further question: how subtle is the relationship between movement and basal rate? One shorebird that has a nearly worldwide distribution is the pied stilt (*Himantopus hemantopus*). A distinct subspecies (*H. h. leucocephalus*) is found in Australia and New Zealand, although it may have arrived in New Zealand from Australia as recently as the early 1800s (Heather & Robertson 1966) as a breeding species. Does the local population in New Zealand have an intermediate or low basal rate, if it is restricted to these islands? If so, would it be as low as might be found in the black stilt (*H. novaezelandiae*), which is endemic to New Zealand? Hawaii also has a sedentary subspecies of the pied stilt (*H. h. knudseni*), which should be examined.

Jetz et al. (2008) made a general analysis of the relationships of avian migration to climate, phylogeny, and BMR in 135 species, including 71 nonmigrants and 64 migrants. They concluded that these relationships are exceedingly complex because of the interactions of various factors that may influence the use of migration. Migrants generally have higher BMRs than nonmigrants, which is compatible with the observations on New Zealand ducks. A complication is that migration is more prevalent among species that breed in cold environments. This correlation is related to various characteristics of the breeding environment, including average annual temperature, potential average evaporation, and the difference between the average January and July temperatures (in the Northern Hemisphere). When the data were corrected for one or more of these variables, no significant difference was found between migrants and nonmigrants. The average annual temperature was the single most important factor predicting mass-independent BMR, as would be expected because migration is most prevalent in polar breeders. Jetz and colleagues also found no evidence that long-distance migrators had higher BMRs than short-distance migrators. The difficulty with this type of analysis is that many of these factors are often correlated with each other and with latitude, which, however, does not deny their importance. Jetz and colleagues also concluded that "for migratory tendency the phylogenetically labile environmental control of BMR emphasizes the significance of phenotypic plasticity."

The difference in BMR between migratory and sedentary species might reflect a difference in body composition, especially the presence in migratory species of larger pectoral muscles, hearts, or digestive organs that process food intake. Flighted rails have higher BMRs and larger pectoral muscle masses than flightless rails, often of the same genus (see chap. 9). As we have seen, the flight muscles of some shorebirds can be hypertrophied with fat deposits, which fuel migratory expenditures. Much more information on body composition and energy stores in relation to migration is needed from a diversity of species.

Migration is a doubled-edged evasion. It is a means of avoiding seasonally harsh conditions often coupled with a low or vanishing food supply. This evaluation surely applies to most polar and many temperate birds, but it raises a question: why bother to live in an area that must be seasonally abandoned? The answer has several components. (1) Temperate and polar environments have seasonally abundant resources. (2) Life in tropical environments is crowded with many potential or actual competitors, which encourages the occupation of low-diversity environments. (3) Life in low-diversity environments may also reduce predator-based mortality sufficiently to compensate for the mortality associated with migration. In a remarkable study, McKinnon et al. (2010) demonstrated that the survival time of artificial shorebird nests containing quail eggs increased

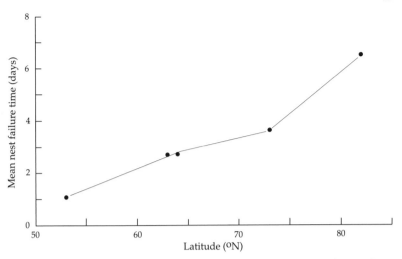

Figure 8.2. Mean time for nest failure along a latitudinal gradient in Canada. (Data from McKinnon et al. 2010.)

over 3,350 km of latitude in Arctic Canada (fig. 8.2). Therefore, migration permits species to exploit seasonally rich resources, evade a congested life in the tropics, increase the probability of successful reproduction, and avoid the seasonally inhospitable environments at temperate and polar latitudes.

A downside must be that an appreciable mortality is associated with spring and fall migration, which in part accounts for the higher rate of reproduction of temperate passerines compared with tropical species. Gavrilov (1998, 1999) made this suggestion and coupled it with the higher basal rates in temperate species. Another implicit downside of migration is its presumptively high cost. However, using doubly labeled water (see box 11.1), Wikelski and collaborators (2003b) measured the cost of migration for up to 600 km and 7.7 hours of movement in two species of thrushes (*Catharus*). Given ambient temperatures encountered similar to those of central Illinois, they found that of the 42 days required to migrate the 4,800 km from Panama to Canada, 18 involved nocturnal flight. The total cost of migration was approximately 4,450 kJ, with 2,340 kJ, or 53%, being the cost of flight. That means that 47% of the cost was during stopovers, a value that depended on stopover temperature. The thrushes did not fly on cold nights; then the cost of a stopover was equivalent to flight for 2.5 hours. Clearly, the cost of migration in these passerines is less than expected. The extent to which these calculations apply to large or long-distance migrators that depend less on stopovers is unknown.

A suspended life

Energy expenditure in a seasonally hostile environment can be markedly reduced by suspending a continuous commitment to a high, regulated body temperature. Such suspensions, however, are time limited. Torpor can be sustained for only a limited period because, even when metabolism is depressed, an endotherm has not given up its body functions, and their maintenance and control must be fueled. In torpor, the regulation of body temperature has not been abandoned, but the maintenance of the body's high normothermic temperature has been modified. Most of the energy used during torpor is derived from stored body fat, although some rodents cache food that is consumed during periods of arousal between periods of torpor. The period over which an energy store will facilitate survival depends on ambient and body temperatures: the greater the temperature differential and the higher the body temperature, the shorter the period over which torpor can be maintained.

Mammals that enter torpor are usually characterized, when regulating body temperature at its normal level, by basal rates that are lower than those found in species of the same mass that do not enter torpor (see chap. 4 and fig. 4.2). The difference in BMR between species that enter torpor, when they are normothermic, and those that do not is greatest at small body masses as a result of conforming, or not, to the "boundary curve for endothermy" (eq. [4.1]).

Torpor comes in two forms, daily and seasonal, which are usually distinguished by length and season. Daily torpor occurs for one or a few days in any season, usually in small species, whereas seasonal torpor can occur either in winter (hibernation) or summer (aestivation), usually for an extended period, in species of various sizes. Complications in these dichotomies exist, however. For example, Arlettaz et al. (2000) observed that the European free-tailed bat (*Tadarida teniotis*) remained torpid for 1 to 8 days in Switzerland, and Turbill & Geiser (2008) and Stawski et al. (2009) found that two species of bats (*Nyctophilus*) were torpid for an average of 4 to 5 days and occasionally up to 10–11 days during short periods of cold weather in subtropical Australia. Both studies referred to this behavior as "hibernation," although it could also be called extended daily torpor. Hibernation, as seen in the Northern Hemisphere, is for an extended period, usually months, reflecting extended periods of low to very low ambient temperatures and a shortage of available resources, conditions usually not found in the Southern Hemisphere, except in southern South America. Both environmental factors are less extreme in the two examples cited above, reflecting a rather mild winter in subtropical Australia and a variable ambient temperature at the southern edge of a temperate environment in Switzerland. During warm spells, the food of these bats, flying insects, is available.

Part of the problem in distinguishing daily torpor from hibernation reflects a gradation of period length between one condition and the other. Daily torpor may be extended during a cold spell, especially when the animals are in poorly sheltered roosts, which was the case in all three species of bats described above, in clefts in exposed rock cliffs in southern Switzerland and under tree bark in Australia. Complications in the definitions of these terms are seen in bears, which avoid demanding conditions in Northern Hemisphere winters by denning, but are not continuously torpid; denned females even give birth to and nurse offspring while in torpor. However, the difference in hibernation between bears and, say, ground squirrels may simply reflect the difference in mass. At the other extreme, many cold-temperate bats that hibernate have modified their reproductive behavior by copulating in the fall, which requires females to store sperm and remain torpid throughout winter. In this case, hibernation is obligatory, at least in females. The difference between seasonal and daily torpor may be arbitrary except in the most extreme cases, just as the difference between daily torpor and the maintenance state is difficult to define clearly (Schleucher 2004).

Daily torpor

Daily torpor, as its name suggests, usually lasts for only a few hours, normally during the night in diurnal species and during the day in nocturnal species. It may occur more frequently in one season than in others, depending on ambient temperature and food availability. As we saw in chapter 4, daily torpor occurs most often in small species, in which it is facilitated by high mass-independent cooling and heating rates, which vary inversely with body mass. The difficulty at a small mass, of course, is the problem of balancing an energy budget with a limited food supply because the starvation time that can be tolerated decreases with a decrease in mass. The occurrence of torpor is most marked in species that feed on foods that are highly variable in availability, including insects (didelphids, dasyurids, shrews, tenrecs, bats, primates, swifts, caprimulgids, hummingbirds, todies, swallows, sunbirds, manakins, white-eyes, titmice), nectar (hummingbirds, sunbirds, manakins), fruit (primates, mousebirds, manakins), and mast crops (rodents, titmice). Daily torpor is frequent in tropical rainforests during periods of heavy rainfall and in scrublands during periods of extended drought, both of which make foraging undependable and food availability limited.

The reduction in energy expenditure associated with the use of daily torpor may be appreciable. For example, a comparison of a 4 g *Sorex cinereus*,

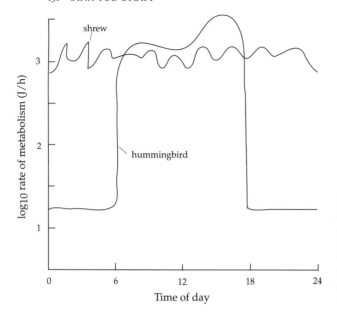

Figure 8.3. Log$_{10}$ rate of metabolism as a function of time of day in a soricine shrew (*Sorex cinereus*) and a hummingbird (*Calypte anna*). (Modified from Bartholomew 1977.)

which does not enter torpor, with a 4 g Anna's hummingbird (*Calypte anna*), which does, indicates that the hummingbird expends only about 50% of the shrew's daily energy expenditure (Bartholomew 1997) (fig. 8.3). The difference in expenditure results from the restriction of the hummingbird's activity to daylight hours and to its entrance into torpor at night.

The time birds spend in daily torpor is incremental. Hiebert (1990) estimated the energy expenditures of the rufous hummingbird (*Selasporus rufus*) at various ambient temperatures as a function of the time spent in torpor (fig. 8.4), as has also been done in a pocket mouse (*Perognathus longimembris*) (fig. 8.5). Furthermore, the time that another pocket mouse (*P. californicus*) spends in torpor varies inversely with the amount of food (seeds) consumed (fig. 8.6). Daily torpor obviously greatly reduces energy expenditure.

BIRDS

Much diversity exists among birds in their use of daily torpor. All hummingbirds that have been studied enter torpor, irrespective of whether they live in a temperate or a tropical environment and whether they are large or small, but the occurrence of torpor is highly variable. Its occurrence is often triggered by a shortage of food, the usual preparatory response being the storage of body fat. For example, Anna's hummingbird (*Calypte anna*), found on the west coast of the United States, is fairly stationary in distribution. It

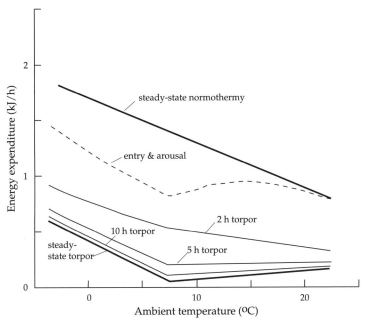

Figure 8.4. Energy expenditure as a function of ambient temperature and time in torpor in a rufous hummingbird (*Selasphorus rufus*). (Modified from Hiebert 1990.)

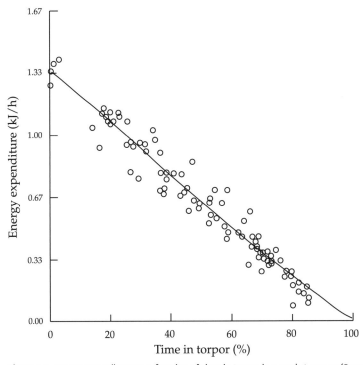

Figure 8.5. Energy expenditure as a function of time in torpor in a pocket mouse (*Perognathus longimembris*). (Modified from French 1976.)

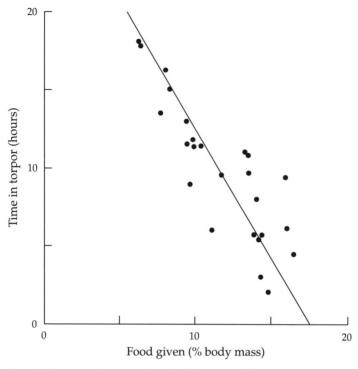

Figure 8.6. Time spent in torpor as a function of the amount of food given to a pocket mouse (*Perognathus californicus*). (Modified from Tucker 1966.)

does not accumulate much fat and only occasionally enters torpor (Beuchat et al. 1979). In contrast, the rufous hummingbird (*Selasphorus rufus*), which is found from Northern California to coastal Alaska and is highly migratory, stores fat during the day and conserves it at night by entering torpor (Beuchat et al. 1979; Carpenter & Hixon 1988). The so-called giant hummingbird (*Patagona gigas*), which weighs 18 to 20 g, is found in Andean South America, usually at altitudes between 2,000 and 4,000 m; this bird also goes into torpor (Lasiewski et al. 1967). Given the huge diversity of hummingbirds, which includes about 328 species that belong to 102 genera (Schuchmann 1999), the vast majority in the Neotropics, much work needs to be done to clarify the conditions and behaviors that promote entrance into torpor. The fawn-breasted brilliant (*Heliodoxa rubinoides*), which is found from 1,800 to 2,500 m in the Andes, demonstrated a BMR that was about 150% of the expected value with only an occasional entrance into torpor (Tellkamp & McNab, personal observation).

One of the clearest descriptions of the ecological context for the occurrence of daily torpor in a tropical bird was a study by Merola-Zwartjes and Ligon (2000) on the Puerto Rican tody, *Todus mexicanus* (misnamed because all todies are limited to Caribbean islands; this species is found only on Puerto Rico). Todies are predominantly insectivorous. Given its mean mass (6.3 g), this species has a basal rate during the nonbreeding season that is 85% of the value expected for birds generally, but 122% of that for nonpasserines. A better standard for its basal rate would be the boundary curve for endothermy (eq. [4.1]); its basal rate is only 72% of that value (see fig. 4.7), so it is not surprising that this species enters torpor. Like all coraciiforms, this species gets some protection by nesting in cavities, although at night it roosts singly in trees. This tody faces high ambient temperatures in dry scrub vegetation at low altitudes (75 m, 750 mm rainfall) and cooler temperatures in rainforests at higher altitudes (400 m, 3,000 mm rainfall). During the breeding season, the tody has a basal rate equal to 73% of the value expected for birds at the lower altitude and 99% at the higher site. Torpor, however, was observed only in females during the breeding season. When in torpor at an ambient temperature of 15°C, body temperature fell to 23.5°C–29.3°C, and the resting rate of metabolism was 31% of the normothermic rate.

Some frugivorous birds enter torpor, including the Neotropical manakins (Pipridae, Passeriformes) and the African mousebirds (Coliidae, Coliiformes). Two manakins, *Manacus vitellensis* (12 g) and *Pipra mentalis* (16 g), have basal rates that are 94% and 97% of the value expected for birds (Bartholomew et al. 1983), which are 99% and 110% of the values expected from the boundary curve—that is, just at the edge (see fig. 4.7)—but they enter torpor. At night, their body temperatures fell as low as 27°C. Under these circumstances, the energy expenditure of manakins equals 42% of that associated with normothermia. A similar reduction of rate of metabolism and entrance into torpor was seen in some mousebirds (Bartholomew & Trost 1970; Prinzinger et al. 1981; Hoffmann & Prinzinger 1984; McKechnie & Lovegrove 2001; McKechnie et al. 2004), only some of which have a tropical distribution and all of which have masses that fall between 35 and 70 g.

The association of daily torpor in birds with food availability is most clearly seen in species that feed on flying insects, a temporally unreliable food supply. They include temperate and tropical goatsuckers (Caprimulgiformes), swifts (Apodiformes), and swallows (Hirundinidae, Passeriformes). Torpor has not been described in Neotropical flycatchers (Tyrannidae), some of the small species of which (e.g., *Mionectes oleaginous* and *Empidonax virescens*) marginally fall below the boundary curve (see fig. 4.7).

Schleucher (2004) reviewed the occurrence of torpor and its correlations in birds. In that analysis, the factors that appeared to influence the presence of torpor included energy shortage, body mass, diet, foraging strategy, and phylogeny. The foods associated with torpor were fruit, nectar, and insects. Furthermore, species that used hovering flight and aerial insectivory also tended to use torpor. Schleucher pointed out the difficulty associated with the definition of terms used in the analysis of torpor.

MAMMALS

As noted in chapter 4, some shrews—namely, species belonging to the genera *Crocidura* and *Suncus*, which belong to the Crocidurinae—readily enter daily torpor. Whether their torpor can be extended for a few consecutive days during a period of harsh weather is unclear. The Crocidurinae are found in southern Europe, southern Asia, and Africa in regions that generally lack extended cold winters. The data available on the energetics of tropical crocidurines indicate that they have basal rates similar to those of temperate species (Sparti 1990). The Soricinae, in contrast, are found in Europe, in northern Asia, and in North America south through the lowlands and mountains of Central America to the mountains of northern South America. *Notiosorex crawfordi*, a small shrew (4.0 g) from the desert Southwest of North America, is the only soricine known to enter torpor. The use of torpor by this species appears to be a response to a combination of a small mass, a diet of soil invertebrates, and life in a desert environment.

Why should very small mammals in the coldest environments be committed to continuous endothermy while their relatives that live in warmer climates use torpor? The standard basal rates of crocidurines were explained by Sparti (1990) as reflecting the warm environmental conditions in the tropics, which may explain the rates in their derived temperate representatives. As a result of these rates and their very small body masses, crocidurines enter torpor because they have not increased their rate of metabolism sufficiently to avoid entrance into torpor. A third soricid subfamily (Myosoricinae) exists; it includes only three genera, which are limited in distribution to Africa. One member of this subfamily, *Myosorex varius*, has a variable body temperature and a basal rate that is 112% of the mammalian standard at a mass of 12.3 g (Brown et al. 1997), a value that is only 68% of that expected from the boundary curve (see fig. 4.1). The persistence of endothermy in some of the smallest *Sorex* in the coldest environments remains unexplained, unless its permits them to take reproductive advantage of resources should they become available (see chap. 14).

The only genus of shrews in South America is *Cryptotis*, a member of the Soricinae. It is found principally in Colombia, Ecuador, and Venezuela at altitudes between 2,500 and 4,000 m. None of the South or Central American members of this genus have had their energetics examined. The only representative of this genus in North America, *C. parva*, has been examined (McNab 1991). It has, for soricine shrews, a rather low mass-independent basal rate of 146%, which is only about 50% of the values found in similarly sized *Sorex* and 67% of the value expected from the boundary curve (see fig. 4.1), and is similar to that of some crocidurines of the same mass (6.2 g). However, no evidence of torpor has been found in this species. Its body temperature was 37.0°C at ambient temperatures between 5°C and 33°C. However, Layne and Redmond (1959) measured a mean core temperature in *C. parva* equal to 35.0°C, with individual measurements as low as 31.9°C., which might indicate a propensity to enter torpor under the appropriate conditions. A close examination of the energetics of *Cryptotis* from Central and South America, especially at low altitudes in Mexico and Panama as well as at high altitudes in South America, may help define the conditions that are required for the evolution of daily torpor in shrews.

As is the case in birds that feed on flying insects, daily torpor is common in many insectivorous bats, presumably reflecting the combined influences of their small mass and short-term variation in their food supply. As a consequence, these bats have low basal rates of metabolism and a propensity to enter torpor even at moderate ambient temperatures. This pattern is most clearly seen in species that belong to the Vespertilionidae and, to a lesser extent, Molossidae, even among tropical species (Leitner 1966). Little evidence of the use of torpor has been found in species that belong to strictly tropical families, including Hipposideridae, Emballonuridae, and Mormoopidae, although these bats show some imprecision of temperature regulation (Bonaccorso & McNab 2003). Schleucher suggested that the best criterion to distinguish between daily torpor and normothermia is whether a species' rate of metabolism at cool ambient temperatures falls below the basal rate of metabolism.

Of special interest are members of the Molossidae. This group has a worldwide tropical distribution, but some species marginally enter warm-temperate environments. As we have seen, *Tadarida teniotis* is found in southern Europe; *T. brasiliensis* and *Eumops perotis* in the southern United States; *T. brasiliensis* in central Chile and Argentina; *T. australis* in temperate Australia; and *E. bonariensis* in central Argentina (see chap. 12). The response of molossids to winter is variable. In North America, *T. brasiliensis* goes into daily torpor during cool weather (Herreid 1963), but remains active throughout the year in the southeastern states, Oregon, and California, where ambient temperatures are usually warm and food remains available. However, on the Great Plains and in the Southwest, this species migrates to Mexico to avoid cold winters (Cockrum 1969; Glass 1982). The physiology and behavior of *T. teniotis* are similar to those of *T. brasiliensis*, which raises the question of whether the term hibernation applies to either.

Daily torpor is widespread in small rodents. It occurs in most species that weigh less than 10 g and in many that are larger. For example, torpor occurs in small cricetine rodents such as *Baiomys taylori* (7.3 g [Hudson 1965]),

Reithrodontomys megalotis (7.6 g [Thompson 1985]), and several species of
Peromyscus (McNab & Morrison 1963; Gaetner et al. 1973; Hill 1975; Lynch
et al. 1978; Deavers & Hudson 1981; Vogt & Lynch 1982; Tannenbaum &
Pivorun 1984).

Among the Heteromyidae, daily torpor has been found in all pocket mice
studied (genera *Perognathus* and *Chaetodipus*), which range in size from
8.3 to 39.5 g, presumably in relation to their seed-eating life in deserts (see
fig. 7.5). Species belonging to the genus *Liomys*, which live in deserts and
have masses between 40 and 50 g, have basal rates equal to 88% and 95% of
those expected from the standard curve (Hudson & Rummel 1966). Their
basal rates are below the boundary curve, but these spiny mice have shown
no evidence of entrance into torpor. *Heteromys*, a rainforest genus, does
not enter torpor and has basal rates that vary between 121% and 136% of
the standard curve (McNab 1979a; Hinds & MacMillen 1985) and are well
above the boundary curve (see fig. 4.3). As noted in chapter 4, the larger
members of this family, kangaroo rats (*Dipodomys*), which weigh between
38 and 110 g, have basal rates between 90% and 110% of the value expected
for mammals, and only the smallest are known to enter torpor. Clearly,
basal rates and the occurrence of torpor in this family vary with body size
and the environment in which the animals live.

Strangely, few representatives of the Muridae have shown a tendency to
enter torpor, even among species that weigh less than 10 g, including *Micromys minutus* (7.4 g [Grodzinski et al. 1988]) and *Mus minutoides* (8.3 g
[Webb & Skinner 1995]). These two mice have BMRs that equal 145% and
137% of the mammalian standard curve, respectively, but, given their small
masses, are only 72% and 71% of the values expected from the boundary curve (see fig. 4.4). Torpor has been found in some xeric or desert-
distributed murid species, including *Acomys russatus*, *Gerbillus pusillus*, and
Pseudomys hermannsburgensis; these species are characterized by a small
mass, a low BMR, or both (Shkolnik & Borut 1969; MacMillen et al. 1972;
Buffenstein & Jarvis 1985). A difficulty with determining what species use
torpor is that the absence of evidence is not always reliable: some species
may require specific conditions, or a trigger, to enter torpor.

The impact of small size on the occurrence of daily torpor, of course,
is influenced by the availability and nutritional quality of the foods consumed. In this context, no arvicoline rodent is smaller than 15 to 20 g, as if
that is what is required of arvicolines to ensure a commitment to continuous endothermy, given their diet. Thus, the two smallest arvicolines studied, *Microtus subterraneus* (17.8 g) and *Myodes californicus* (18.3 g), have
BMRs that are 177% and 216% of the values expected from mass—which,
given their masses, are high enough to exceed the values expected from
the boundary curve (see fig. 4.1). These arvicolines have made a greater

commitment to endothermy than the two smallest murids that have been studied, given the combination of their higher BMRs and larger masses. The commitment to continuous endothermy by soricine shrews and arvicoline rodents undoubtedly reflects the combined effects of body mass and food, with a smaller mass in species that consume high-protein insects and a larger mass in species that eat high-fiber plants. What is occurring in small murids is unclear.

Seasonal torpor

Seasonal torpor comes in two forms: hibernation in winter and aestivation in summer. Unlike daily torpor, which tends to be limited to small species, mainly because of the rapid turnover of resources and the inability of small animals to store much energy, seasonal torpor can occur in species of all sizes, the largest being black and grizzly bears. Mammals that hibernate include some small marsupials, some primates, many ground squirrels, kangaroo mice, jumping mice, hamsters, many temperate bats, and some carnivores. The poorwill (*Phalaenoptilus nuttallii*) is the only bird suggested to hibernate. Aestivation occurs in some ground squirrels, mice, and tenrecs but is not known to occur in birds.

HIBERNATION

Hibernation is a response to harsh environmental conditions that occurs in mammals living in cold-temperate and polar climates, where ambient temperatures may fall as low as −20°C to −40°C for extended periods. Preparation for hibernation occurs in late summer, after reproduction has occurred, and early fall, when there is significant deposition of body fat in most hibernating species.

As in daily torpor, the rate of metabolism of hibernating endotherms decreases with a decrease in the regulated body temperature. This means that in preparation for hibernation, the animal must find a shelter that will provide an ambient temperature above 0°C, but cool enough to depress metabolism. Within the range of acceptable hibernating temperatures, the period that an individual can remain in torpor depends on the ambient and body temperatures, but it is normally not continuous throughout the hibernation period. If body temperature increases, the rate of metabolism increases (fig. 8.7), nitrogenous wastes increase, water balance is threatened, and a torpid individual will awaken. The duration of a torpor bout varies inversely with the body temperature of the hibernator (fig. 8.8), but also depends on the size of the hibernator (French 1985).

The hibernaculum temperatures chosen vary inversely with body mass

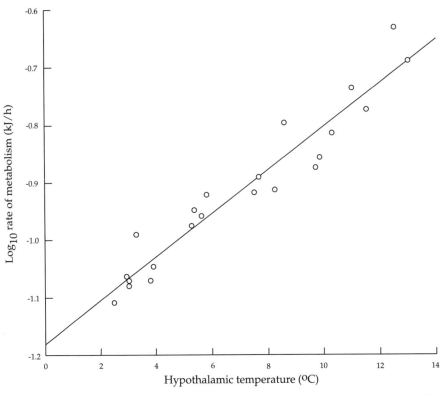

Figure 8.7. Log$_{10}$ rate of metabolism as a function of hypothalamic temperature in a ground squirrel (*Spermophilus lateralis*). (Modified from Hammel et al. 1968.)

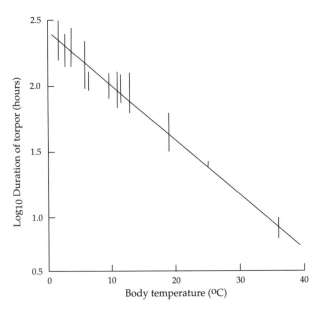

Figure 8.8. Duration of torpor as a function of body temperature in a ground squirrel (*Spermophilus lateralis*). (Modified from Twente & Twente 1965.)

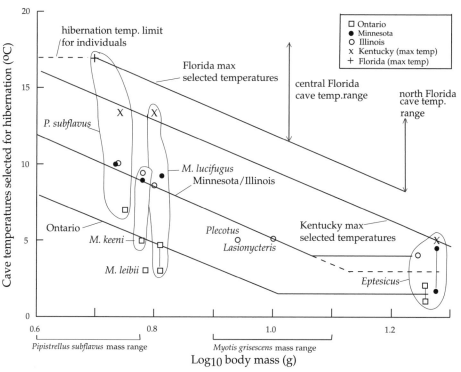

Figure 8.9. Cave temperatures selected for hibernation as a function of $\log_{10}$ body mass in bats in central and southern North America. (Modified from McNab 1974.)

(fig. 8.9). In Florida, a warm-temperate to subtropical peninsula, three species of vespertilionid bats are found in the region's many limestone caves. These three species, *Pipistrellus subflavus, Myotis austroriparius*, and *M. grisescens*, which weigh about 5, 7, and 10 g, respectively, differ in their responses to ambient temperature, which are modified by the presence or absence of clustering behavior (McNab 1974). *Pipistrellus subflavus* is a tree bat that hibernates in caves as solitary individuals, although a sexual pair may be temporarily clumped together. *Myotis grisescens* and *M. austroriparius* are clustering bats.

Myotis grisescens and *P. subflavus* are obligatory hibernators with a reproductive scheme in which copulation occurs in the fall; females must remain torpid to store sperm, which permits ovulation, fertilization, and birth to occur in the spring. The pipistrelle remains torpid in Florida at cave temperatures between 12°C and 17°C by hibernating as solitary individuals and by having a small mass; as a result, ΔT during torpor is usually 0.3°C to 0.6°C (McNab 1974). The impact of cave and core temperatures is shown by the observation that at a cave temperature of 10°C in Kentucky, 50% of the

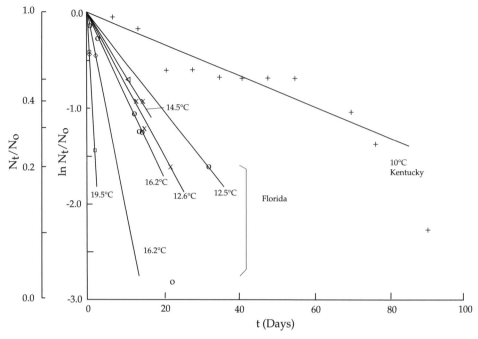

Figure 8.10. Lognormal of the proportion of populations of the bat *Pipistrellus subflavus* remaining in torpor as a function of time and temperature in caves. (Modified from McNab 1974.)

hibernating *Pipistrellus* awakened from torpor and moved in 42 days (Davis 1965), whereas at a cave temperature of 16.2°C in Florida, 50% moved in 7 days, and at 19.5°C, 50% moved in 1 day (fig. 8.10).

This pattern can be quantitatively described by the equation

$$N_t/N_o = e^{-kt}, \tag{8.1}$$

where N_o is the original number of hibernating individuals, N_t is the number remaining in hibernation after a time period t, and k is a rate constant ($1/t$). The rate constant is principally determined by two factors: hibernaculum temperature and body mass. The constant k increases with hibernaculum temperature (fig. 8.11). Clearly, a temperature limit to hibernation exists. That limit depends on body mass because k increases with mass (McNab 1974), which means that large species must select lower hibernaculum temperatures than small species to remain in torpor for the same time period. The highest temperature at which *Pipistrellus* can hibernate is about 17°C (see fig. 8.9).

The gray myotis (*M. grisescens*), in contrast, combines clustering behavior with a comparatively large mass, and therefore requires colder cave temperatures to remain torpid than *Pipistrellus*. As a result, the gray myotis cannot find sufficiently cold temperatures in Florida (see chap. 12). *Myotis austroriparius* does not require cold caves, in spite of its intermediate size and clustering behavior, because this species copulates in spring; a normothermic state is usually acceptable in winter because an appreciable population of flying insects is present throughout the year in Florida. However, this species is also found as far north as southern Illinois, where conditions require it to hibernate. Whether it shows a shift in reproductive behavior is unclear, but the possibility raises some questions. Do all, most, or only some temperate vespertilionids show such a shift? How flexible is reproductive behavior in vespertilionids?

An adequate preparation is required to fund an animal's (reduced) energy expenditure during the hibernation period. This preparation usually takes the form of storing body fat, although some rodents store seeds and nuts. How do bats store fat during the fall when air temperatures are low and the supply of insects is reduced? Krzanowski (1961) suggested that bats choose cool roosts, their body temperature falls, and the cost of body maintenance decreases, which permits them to accumulate body fat. A similar pattern was seen in premigratory birds (Carpenter & Hixon 1988; Dawson & Whittow 2000; Merola-Zwartjes & Ligon 2000; Butler & Woakes 2001; Schleucher 2004).

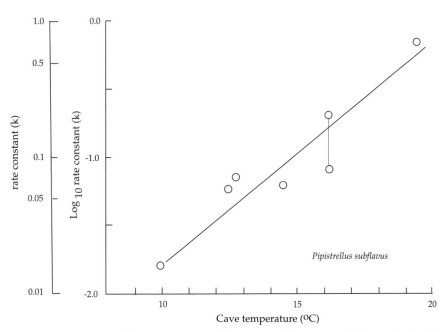

Figure 8.11. $\log_{10}$ of the exponential power of the turnover of *Pipistrellus* populations as a function of cave temperature. (Data from McNab 1974.)

Speakman & Rowland (1999) examined this suggestion in *Plecotus auritus*. This bat preferentially selected cool ambient temperatures, which through its entrance into torpor facilitated an increase in body mass, thereby confirming Krzanowski's suggestion. However, a tropical bat (*Nyctophilus bifax*) showed torpor more frequently if it had large fat deposits (Stawski & Geiser 2010), an observation that reemphasizes the physiological and behavioral difference between tropical and temperate insectivorous bats.

Another question of interest is whether cave-dwelling bats behave differently from tree bats, at least in winter. In cold climates, nonmigratory bats hibernate in caves or buildings, awaking occasionally depending on body temperature. Tree bats either are limited to warm climates or avoid cold winters by migrating to them. Activity in these conditions can be year-round, even if limited to warm periods when flying insects are present. This is undoubtedly the case in *Nyctophilus* and some temperate bats that migrate to warmer climates, including species belonging to the genera *Lasiurus*, *Nycticeius*, and probably *Nyctalus*. These species appear to be more inclined than temperate vespertilionids to maintain a body temperature intermediate between deep torpor and a normothermic condition (Genoud 1993). This behavior would facilitate their capacity to exploit available resources without being committed to a high energy expenditure during a period of food shortage and yet permit them to avoid damage during a short period of low ambient temperatures.

Recent observations have demonstrated hibernation in a tropical primate, the fat-tailed dwarf lemur (*Cheirogaleus medius*) from Madagascar (Dausmann et al. 2004). This fruit- and nectar-feeding primate may hibernate for up to 7 months. Its body temperature usually varied from 15°C to 30°C, reflecting the variation in the temperature of the tree hole in which it was located; the mean ΔT equaled 1.8°C. The variation in the temperature of the hibernaculum and in the lemur depended on the thermal properties of the tree in which the lemur was found.

The use of hibernation has many liabilities, including helplessness against a predator, the cessation of growth, and a reduction of reproductive output, with a consequent reduction in population r_{max}. These restrictions may be the reason why relatively few mammals hibernate: they may require extreme environmental conditions to withstand competition from trophically similar non-hibernators. These factors may also explain why birds preferentially migrate in response to harsh environmental conditions and why hibernation is almost unknown in birds.

The one probable example of hibernation in birds is found in a caprimulgid, the poorwill. Jaeger (1948, 1949) found a torpid poorwill in the same niche in a California canyon wall in three consecutive winters. He estimated that the poorwill was torpid for about 85 days in 1947–1949 and lost 8.1 g during a 41-day period. Later field studies of the poorwill were made by Brigham (1992) and Firman et al. (1993) in Canada, where it occurs only in summer. They showed that daily torpor regularly occurred in spring and fall, but not during the breeding season. This, however, does not answer the question of whether the poorwill goes into hibernation in its winter range in the desert Southwest and northern Mexico, and if so, for how long. The complication is that many caprimulgiforms go into daily torpor, including other caprimulgids (*Caprimulgus europaeus* [Peiponen 1965, 1966, 1970], *C. vociferous* [Lane et al. 2004], *C. tristigma* [Mckechnie et al. 2007], *Chordeiles acutipennis* [Marshall 1955], and *Ch. minor* [Lasiewski & Dawson 1964]); a eurostopodid (*Eurostopodus argus* [Dawson & Fisher 1969]); and an owlet-nightjar (*Aegotheles cristatus* [Brigham et al. 2000]), presumably in association with an undependable diet of flying insects. In the case of the owlet-nightjar, daily torpor occurs in winter and lasts up to 9 hours during the day, while the bird is roosting in caves and feeding at night. Körtner et al. (2001) found that seven individuals of the large (500 g) tawny frogmouth (*Podargus strigoides*) entered torpor on 44% of the 462 days of observation, often twice a day, for about 3.5 hours near dawn and 7.0 hours at night. Passive rewarming during the day brought them out of torpor.

How can one distinguish between sequential daily torpor and hibernation, or is the difference between these states completely arbitrary? Bears have been thought to be marginal hibernators, even though they shelter in dens for up to 5–7 months, principally because their body temperatures do not fall to low values. Furthermore, the birth of young occurs while the female is hibernating. The rationale for this timing is probably that the combined period of gestation and growth of young to independence must occur before the female must prepare for hibernation in the next fall. Studies of black bears (*Ursus americanus*) have measured body temperatures as low as 30°C to 31°C (Hock 1960; Tøien et al. 2011). The peculiar aspect of this shallow hibernation is that it leads to a 50% (Hock 1960) to 75% (Tøien et al. 2011) reduction in metabolism compared with the normothermic rate. This reduction in metabolism is much greater than expected from the limited reduction in body temperature, which suggests that some other factors are involved. In fact, black bears retain a low rate of metabolism for up to 3 weeks after emerging from their dens (Tøien et al. 2011). Female polar bears (*U. maritimus*) also den in winter, during which time body temperature falls to 36°C and rate of metabolism to 50%–70% of their normal rate (Watts & Hansen 1987).

A similar reduction in metabolism beyond that expected from a decrease in body temperature was found in the tropical binturong (*Arctictis binturong*) (McNab 1995). At ambient temperatures <15°C, rate of metabolism fell by about 50% with a fall in body temperature of about 2°C (see fig. 1.8).

A somewhat similar reduction in rate of metabolism was found in the African palm-civet (*Nandinia binotata*). Again, some factor other than the fall in body temperature must account for the decrease in rate of metabolism in the binturong. In the case of the binturong and the palm-civet, this behavior is not hibernation, but a response to low ambient temperature, the ecological significance of which is unclear, given their tropical distributions.

Two species of mammals are considered to be marginal hibernators: badgers and beavers. North American badgers (*Taxidea taxus*) sometimes enter a lethargic state with starvation, during which body temperature can fall from 37°C to 28°C (Harlow 1981a,b)—a greater decrease than found in bears, reflecting the badger's smaller size (8 kg). During this period, the rate of metabolism falls about 27%. Torpor also occurs in the European badger (*Meles meles* [Slonin 1952; Johansson 1957]). But whether this torpor is hibernation is uncertain.

Some confusion exists on the question of the response of beavers (*Castor canadensis*) to a cold winter. Beavers are not able to store enough food to maintain their normal level of energy expenditure while overwintering (Novakowski 1967). In one laboratory study, beavers from Yukon that were held in continuous darkness in winter reduced their food intake by 40% (Aleksiuk & Cowan 1969a,b); their thyroid activity fell, their growth stopped, and they became paralytic, but body temperature fell only 1°C. However, Dyck and MacArthur (1992, 1993) were not able to find such responses either in the field in southeastern Manitoba or in the laboratory. They agreed that food caches were often inadequate and suggested that the caches may have been supplemented by feeding on aquatic vegetation. Beavers from California did not show these responses. The responses of beavers may reflect the particular winter conditions encountered.

That some mammals of an intermediate or large size have a physiological state between a complete commitment to hibernation and normal homeothermy should not be surprising, given the diversity of the environments occupied by mammals and their diverse behaviors. Not all endotherms conform to the sharp dichotomy invented by people, and those of intermediate size may be the species that have the greatest freedom to evolve a variety of responses to external conditions. Small species are trapped by the high cost of commitment to continuous endothermy, and large species have masses that prevent rapid cooling and heating of the body, as well as having to tolerate a high equilibrial temperature if in torpor.

AESTIVATION

Aestivation in mammals has attracted little attention. It has been described in a few mammals, including some small rodents (*Baiomys taylori* [Hudson

1965], *Peromyscus eremicus* [MacMillen 1965], *Perognathus longimembris* [Bartholomew & Cade 1957], *Perognathus californicus* [Tucker 1965, 1966], *Spermophilus mohavensis* [Bartholomew & Hudson 1960], *S. columbianus* [Shaw 1925], *S. richardsonii* [Wang 1979], and *S. tereticaudus* [Walker et al. 1979]) and tenrecs (*Geogale, Microgale* [Stephenson & Racey 1993a,b]). The factors that induce aestivation may include food or water shortage. Essentially, aestivation is high-temperature torpor; for example, *P. eremicus* cannot spontaneously arouse from a body temperature less than 16°C (MacMillen 1965).

Some ground squirrels are committed to life in such demanding environments that they enter hibernation in the fall, remain in torpor through the winter, awaken in spring, reproduce, and then in midsummer, during a dry period, begin aestivation, which directly shifts into hibernation in the fall. Torpor in *Spermophilus richardsonii* lasts 8.5 to 9 months on the demanding prairies of Alberta, Canada, so the animals are active for only 25% to 30% of the year. Energy expenditure in this species is about 34% of the amount that would be spent if the animals were continuously normothermic for the entire year and only 12% during the 8 months of torpor (Wang 1979). The Mohave ground squirrel (*S. mohavensis*) is active for about 140 days, 38% of the year (Bartholomew & Hudson 1960), slightly longer than Richardson's ground squirrel because of a less extreme winter, but it faces a more extreme regime of high temperature and low water availability in summer.

CHAPTER NINE. *Island Life*

Endotherms endemic to islands face conditions that are distinctly different from those encountered on continents. These conditions include a restricted geographic area with a resulting limited resource base, low species
diversity, and the absence of eutherian predators (thus this definition of an
"island" almost includes Australia). When continental species immigrate
to islands and establish permanent populations, they tend to conform to a
pattern: small species increase in size, whereas large species decrease in size
(Foster 1964). An increase in size on islands has been seen in rodents (Alder
& Levans 1994), shrews (White & Searle 2007), and many other small mammals (Lomolino 1985, 2005). The shrinking of large mammals on islands
includes deer on Jersey (Lister 1989), hippopotami on Madagascar (Stuenes
1989), elephants on Malta, Corfu, and Cyprus (Sondaar 1977; Roth 1990),
bovids on Majorca (Köhler and Moyà-Solà 2009; Köhler 2010), and mammoths on Wrangel (Vartanyan et al. 1993; Lister 1993). This dual pattern
of change in size is referred to as the "island rule" (Van Valen 1973). The
energy expenditure of island endemics is modified, at least as a result of
changes in body size.

The island rule

The changes in body mass on islands reflect limited availability of resources
and limited community diversity (Foster 1964; Van Valen 1973; Lomolino
2005; McNab 2002b, 2010). If small continental species increase in mass
and large continental species decrease in mass when they colonize islands,
where is the breaking point between small and large species? According to
Lomolino (2005), it depends on the species' ecological niche: it is about 26 g
in shrews, 270 g in rodents, 375 g in carnivores, and 7 kg in artiodactyls.
Meiri et al. (2008a), however, argued that there is no island rule relative
to body size because all these changes in size reflect phylogeny. This argument is an "explanation" for everything and nothing at the same time (see

chap. 2); besides, what happened to resources (Raia & Meiri 2006; Meiri et al. 2008b)? The effect of phylogeny is difficult to distinguish from that of body size because of the correlation of body size with phylogeny: there are no shrews the size of artiodactyls! Phylogeny can be an effective basis for the analysis of character states only if the functional basis for its influence is determined, as in the case of the difference in energetics between marsupials and eutherians (see chap. 14).

Regardless of the "cause" of changes in body size on islands, these changes influence energy expenditures and resource requirements. A decrease in body size reduces resource requirements, which in a resource-limited environment—limited at least by island area—facilitates long-term survival. Small species, on the other hand, can increase their energy expenditure in a diversity-poor environment because they must share resources with fewer species and therefore are unlikely to be resource limited.

Some exceptions to the island rule further the point that resources are the fundamental determinant of body size on islands. For example, the large size of brown bears (*Ursus arctos middendorffi*) on Kodiak Island correlates with a huge seasonally available protein and fat source in the form of five species of breeding salmon (*Oncorhynchus*), a food that permits a large mass in bears (Hilderbrand et al. 1999; Swenson et al. 2007). Arctic foxes (*Alopex lagopus*) are larger on Mednyi Island (Commander Islands, Bering Sea), where they have access to marine vertebrate and invertebrate food resources, than on the Arctic mainland, where they depend on highly variable vole populations (Goltsman et al. 2005). Red foxes (*Vulpes vulpes*) in Scotland are larger where they feed on hares and rabbits, but smaller where they feed on voles; the differential occurrence of these prey depends on local conditions (Kolb 1978). Raia and Meiri earlier (2006) concluded that carnivore size on islands is "influenced by resources and little else." And on large islands, in the absence of large herbivorous mammals, some birds became very large grazers and browsers, including moas (Emeidae, Dinornithidae) in New Zealand and elephant birds (Aepyornithidae) in Madagascar. The island rule, Bergmann's rule, and Dehnel's phenomenon, as well as Cope's rule, which suggests that mammalian size tends to increase with time, appear to be expressions of the same pattern (McNab 2010): mammals increase or decrease in body size with respect to geography or time depending on the abundance, availability, and size of resources, which may be influenced by the necessity to share resources with other species. On the other hand, a resource-poor island may require a reduction in mammalian size beyond what would be expected on a large island, as appears to be the case on Borneo (Meiri et al. 2008b). This pattern was described as a "resource rule" (see chap. 6).

Some birds in the South Pacific have a distinctive distributional pattern.

The island thrush (*Turdus poliocephalus*) is found on a variety of islands from Vanuatu through the Solomon Islands, New Guinea, and the Northern Marianas north to Borneo, the Philippines, and Taiwan. The altitude above which the thrush is found depends on island size and topography. It is found at sea level on Vanuatu and at progressively higher altitudes on larger islands. It is generally found above 1,400 m in the Solomon Islands, between 2,100 and 3,200 m on Borneo, between 2,500 and 4,250 m in New Guinea, at 1,100 – 1,650 m in the Philippines, and at 1,800 – 2,500 m on Taiwan (Collar 2005). This species is not selecting similar microclimates over a range in latitudes. Bird communities are less diverse on small islands, where the island thrush is found at sea level, but as island size increases, lowland bird community diversity increases, and the thrush is restricted to communities at increasingly higher altitudes, where species diversity declines. Mayr and Diamond (2001) suggested that the thrush is selecting communities with lower bird diversities, although Steadman (2006) argued that its absence from lowland communities more likely reflects a human impact. The island thrush at 2,860 m in New Guinea has a basal rate equal to 123% of the value expected of birds generally (McNab, personal observations), which raises the question whether lowland (small island) populations have different BMRs (see chapter 9).

Measurements of the energetics of this thrush from small-island populations would be of special interest. (It is noteworthy that this species shows a great diversity in color pattern with respect to distribution; for example, it is essentially all dark gray in southeastern New Guinea, has a red head on Fiji, and has a white head on Vanuatu. In fact, there are 49 subspecies, which indicates that island populations are isolated from one another. Is that diversity reflected in basal rate as a function of altitude?)

A similar geographic distribution is found in other species. For example, the bronze ground-dove (*Gallicolumbia beccarii*) is found between 1,200 and 2,900 m in New Guinea, near sea level on smaller, nearby islands, and at intermediate altitudes on the larger islands in the Solomon Islands and the Bismarck Archipelago (Baptista et al. 1997). At 2,000 m, this ground-dove has a BMR that is 84% of the value expected for birds (McNab, personal observation). The blue-faced parrot-finch (*Erythrura trichroa*) has a somewhat similar distribution. At an altitude of 2,100 m, it has a basal rate of 101% of the value expected for birds (B. K. McNab, personal observation).

Islander energetics

Many island endemics reduce energy expenditure. This reduction can occur in two ways: through a reduction in body size and through a decrease in mass-independent rates of metabolism. A reduction in the body size of

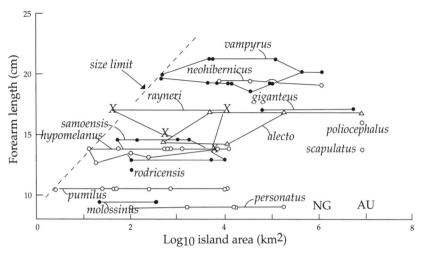

Figure 9.1. Forearm length in flying foxes (Pteropodidae) as a function of log$_{10}$ landmass area in the South Pacific. Abbreviations: NG, New Guinea; Au, Australia. (Modified from McNab 1994b.)

island endemics is seen in rails (Rallidae), flying foxes (Pteropodidae), and fruit pigeons (Columbidae). These reductions can occur both intra- and interspecifically. Individuals of some species are smaller on small islands than they are on larger islands, a pattern seen in rails (McNab 1994b), and some species on small islands are smaller than their relatives on larger islands and continents. For example, the largest flying foxes are found on intermediate to large islands, and those found on the smallest islands are the smallest species (fig. 9.1).

Some endotherms also respond to island life by reducing their mass-independent rates of energy expenditure. This strategy has been found in birds and bats that are restricted to some of the smallest, most isolated oceanic islands, birds and bats being the only endotherms that have the capacity to disperse to these environments. Thus, small-island endemic pteropodids have lower mass-independent BMRs than large island or continental species (McNab & Bonaccorso 2001) (fig. 9.2).

A similar pattern is found in fruit pigeons belonging to the South Pacific genus *Ducula*, some species of which are endemic to the large island of New Guinea; others are limited to the intermediate islands of New Britain, New Ireland, and Bougainville; and still others occupy small islands such as Karkar, Missau, and Manus. Species of *Ducula* that are small-island endemics— namely, *D. pacifica* and *D. pistrinaria*—have low mass-independent basal rates; those from intermediate islands, including *D. bicolor* and *D. rubricera*, have intermediate basal rates, and those endemic to New Guinea, *D. rufigaster*, *D. pinon*, and *D. zoeae*, have higher basal rates (fig. 9.3) (McNab 2000b).

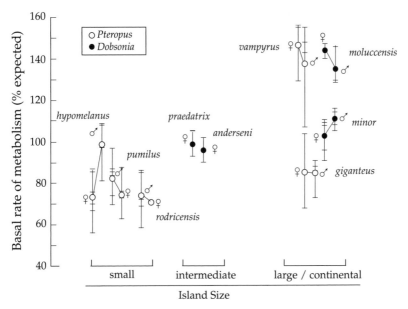

Figure 9.2. Mass-independent basal rate of metabolism in flying foxes as a function of island size. (Modified from McNab & Bonaccorso 2001.)

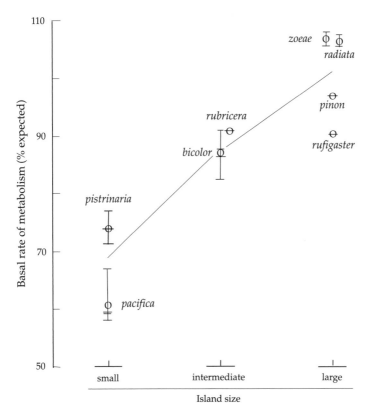

Figure 9.3. Mass-independent basal rate of metabolism in fruit pigeons of the genus *Ducula* as a function of island size in the South Pacific. Horizontal bars indicate the mean BMR of individuals; the circles are species means. (Modified from McNab 2000b.)

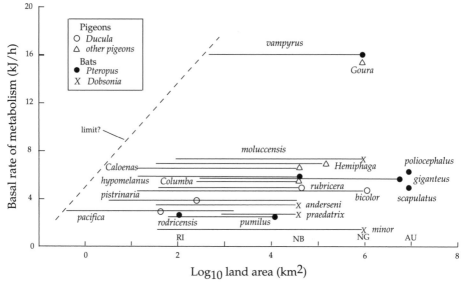

Figure 9.4. Basal rate of metabolism in pigeons and flying foxes as a function of $\log_{10}$ landmass area in the South Pacific. Abbreviations: RI, Rodrigues Island; NB, New Britain; NG, New Guinea; AU, Australia. (Data from McNab 1994b, 2000b, 2002b.)

As a result of the correlation of both body size and BMR independent of body size with island size, a positive correlation is found between the BMR of flying foxes and pigeons in the South Pacific and minimal island size. The data presented in figure 9.4 are the most extensive available on the physiology of endotherms as a function of island size. Much more information is required to determine the extent to which this pattern is widespread or limited to a few species or species groups. Do passerines endemic to small oceanic islands also reduce mass-independent energy expenditure, or does such a correlation require a large body mass? Of course, by virtue of being small, passerines may be able to tolerate life on quite small islands without making any mass-independent adjustments. If passerines respond to an island existence, how small must an island be for a 10–20 g passerine to reduce its mass-independent energy expenditure?

Measurements on some birds from New Zealand, a comparatively large island archipelago, are relevant to these questions. Five species of parrots (Psittacidae) are endemic to this archipelago. They include the large, flightless kakapo (*Strigops habroptilus*), the intermediate-sized kea (*Nestor notabilis*) and kaka (*N. meridionalis*), and three parakeets, yellow-crowned (*Cyanoramphus auriceps*), Malherbe's (*C. malherbi*), and Antipodes Island (*C. unicolor*). Also present is the native red-crowned parakeet (*C. novaeze-*

landiae). The two *Nestor* and the yellow-crowned and red-crowned para-keets have had their basal rates measured (McNab & Salisbury 1995). These measurements are of interest because they relate to the relationships of mass-independent BMR and the size of birds to island size. Here we have intermediate to small species on large islands. The two *Nestor* have basal rates that are 113% and 137% of the values expected from mass, and the two *Cyanoramphus* have basal rates equal to 92% and 112%. These data suggest that with islands as large as New Zealand, no adjustments in mass-independent BMR need to be made, as is the case, of course, on the huge, nearly continental island of New Guinea. Measurements on the energetics of the kakapo would be very interesting, but if it has a low BMR, it would probably be related to its flightless condition, not to its presence on a large island archipelago.

Mathias et al. (2004) measured the resting energy expenditure of house mice (*Mus musculus domesticus*) on the island of Porto Santo, a 45 km² is-land in the Madeira archipelago. These mice were transported to the island some five hundred years ago. Their mass-independent energy expenditure was 64% of that of a population of house mice from continental Portugal. The low energy expenditure of the Porto Santo mice may be associated with their island existence, but is more likely associated with the xeric conditions on the island. What is most interesting is that the response of the island mice to their isolation occurred in such a short time.

The evolution of flightless birds

One of the most striking responses by birds to life on oceanic islands is the evolution of a flightless condition. Flightlessness evolved on islands repeat-edly in the Rallidae (rails and gallinules on hundreds of islands) and the Anatidae (ducks, New Zealand, Amsterdam, southern South America) and occasionally in the Phalacrocoracidae (cormorants, Galápagos), Threskior-nithidae (ibis, Réunion), Rhynochetidae (kagu, New Caledonia), Colum-bidae (dodo, Mauritius; solitaire, Rodrigues), Psittacidae (kakapo, New Zealand), Strigidae (owl, Andros, Bahamas), Acanthisittidae (wrens, New Zealand), and Emberizidae (bunting, Canary Islands), as well as the Aptery-gidae (kiwis, New Zealand). Flightlessness is also characteristic of penguins (Spheniscidae), which might have evolved on islands in the subantarctic, or maybe Antarctica should be considered an ice island (no terrestrial preda-tors are there). Flightless birds are also found in Australia (cassowaries [Ca-suariidae, which are also present in New Guinea], and the emu [*Dromaius novaehollandiae*]); in South America (rheas [Rheidae]); and in Africa and formerly Asia Minor (the ostrich [*Struthio camelus*]). The evolution of a flightless condition, in fact, has occurred many times: Steadman (2006) estimated that the number of flightless rails in the South Pacific was between 444 and 1,579—that is, repeatedly on almost all Pacific islands.

Lakes are similar to islands in that they are limited areas circumscribed by territories that are hostile to aquatic residents, just as the ocean is hostile to terrestrial vertebrates on islands. One difference, however, is that lakes are more subject to aquatic predators than islands are to terrestrial predators. One group of obligate aquatic birds has repeatedly evolved a flightless condition on lakes: grebes. These birds include the Atitlán grebe (*Podilymbus gigas*) in Guatemala, Junín grebe (*Podiceps taczanowski*) in Peru, and Titicaca grebe (*Rollandia microptera*) in Peru and Bolivia. The Junín grebe is apparently extinct, or nearly so, and the Atitlán grebe is extinct. The vulnerability of flightless grebes is illustrated by the extinction of the Atitlán grebe, which was precipitated by the purposeful introduction of bass (*Micropterus*) into the lake, the reason for which was to supply the local human population with a food source!

Why does a flightless condition evolve? An analysis of the energetics of sixteen populations of fourteen species of rails (McNab & Ellis 2006) indi-

BOX 9.1. ·

Searching for a Flightless Rail

As soon as I arrived in Papua New Guinea, I started to ask about the presence of the New Guinea flightless rail (*Megacrex inepta*). I had hoped to compare this species with the two flightless rails of New Zealand and the Guam flightless rails (*Gallirallus owstoni*) that I had borrowed from the Lowry Park Zoo in Tampa. Wherever I went in PNG, I asked about the flightless rail, but everyone said that it was very secretive and I would never get it, and that it might be extinct. Finally, in 1996, while in the National Museum and Art Gallery in Port Moresby, I met a PNG national, Samuel Kepuknai, from Kiunga, a town on the Fly River, and asked him my standard question whether he ever sees the flightless rail, which is supposed to be in the Fly and Sepik river drainages. Samuel stated, "I have two under my house." (Well, yes, I thought, along with the pterosaurs!) But the longer I talked to Samuel, the more I started to believe that he indeed had flightless rails. I made arrangements to have the rails flown to Port Moresby. The birds were in good shape, although one individual had only one leg as a result of being captured with a bow and arrow. In 1997, Jack Kaufman and I went for five days with Samuel to his field station on a tributary of the Fly River, the Elevala River, where we saw freshwater crocodiles, southern crowned pigeons (*Goura scheepsmakeri*), Pesquet's (vulturine) parrots (*Psittrichas fulgidus*), crested hawks (*Aviceda subcristata*), the twelve-wired bird of paradise (*Seleucidis melanoleuca*), many fruit pigeons and doves, and in the early morning, heard the calls of the flightless rail. I am always amazed at how capricious field research in biology is—persistence pays off.

· ·

cated that the ten flighted populations have mass-independent basal rates that average 1.38 times those of the six flightless species (fig. 9.5A). Furthermore, rails can be divided into two groups based on their food habits: herbivorous gallinules and omnivorous rails (fig. 9.5B). The herbivorous species had basal rates that were 1.37 times those of the omnivorous species. Therefore, flighted gallinules have mean basal rates that are 1.38 × 1.37 = 1.89 times those of flightless rails. This analysis is described by the following equation:

$$\text{BMR (kJ/h)} = 329.5 \, (F \cdot V) \, m^{0.631}, \tag{9.1}$$

where F is the dimensionless coefficient for food habits and V is the dimensionless coefficient for a flighted or flightless condition. This relationship accounted for 96.2% of the variation in rallid BMR (fig. 9.6). However, two species are outliers: the New Guinea flightless rail (*Megacrex inepta*) (see box 9.1), whose BMR is lower than estimated by equation (9.1), and the takahe (*Porphyrio mantelli*), whose BMR is quite high (fig. 9.6). These two species are, respectively, the largest omnivore (0.9 kg) and the largest herbivore and largest rail (2.8 kg). Why these two are deviant from the analysis is unclear, except that the takahe lives in the mountains of South Island, New Zealand, where the cool environment might possibly account for its high basal rate. That is, factors other than body mass, food habits, and flight capacity may influence rail BMRs.

Another correlate of a flightless condition and a low mass-independent BMR in this family is a reduction in clutch size to about one-fourth of that in flighted species (fig. 9.7). So, in spite of the usual view that rails are reluctant to fly, they arrive at all oceanic islands, regardless of the island's isolation, possibly aided by storm fronts. One of the first adjustments that they make to island life is the evolution of a flightless condition, often in association with a reduction in body size. However, the factor responsible for the small clutch size is most likely to be a low mass-independent BMR (see chap. 14).

Rails on some islands may respond to a shortage of resources, but their flightless condition has evolved repeatedly on large islands such as New Zealand, New Guinea, and Hawaii. A shortage of resources may not be the fundamental factor pushing rails to a flightless condition, except on the smallest islands. It may be that rails are (phylogenetically, i.e., historically) programmed to be stealthy in cluttered environments, such as swamps, which may facilitate the evolution of a flightless condition on islands, and that eutherian predators are the factor forcing continental rails to retain the capacity for flight. Nevertheless, the occurrence of a flightless condition cannot be predicted by phylogeny because its evolution depends on the accidental arrival of flighted species on oceanic islands.

Does this mean that a flightless condition in birds evolves only in the

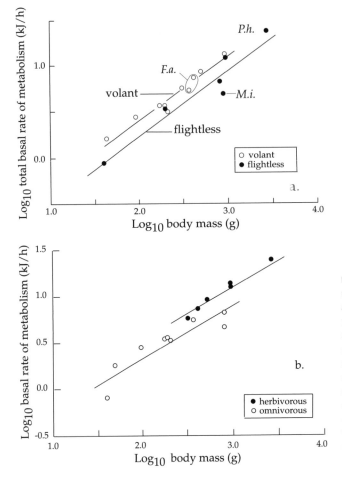

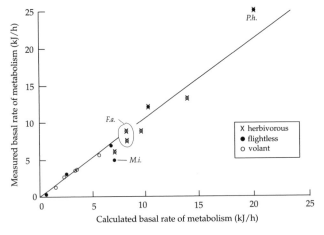

Figure 9.5. $\log_{10}$ basal rate of metabolism as a function of $\log_{10}$ body mass in (A) flighted and flightless rails and (B) herbivorous and omnivorous rails (Rallidae). Abbreviations: *F.a.*, coot (*Fulica atra*); *M.i.*, New Guinea flightless rail (*Megacrex inepta*); *P.h.*, takahe (*Porphyrio mantelli*). (Modified from McNab & Ellis 2006.)

Figure 9.6. Measured basal rate of metabolism in rails as a function of the basal rate calculated from equation (9.1).

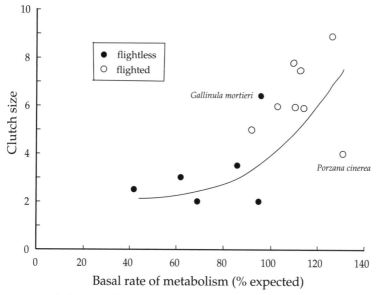

Figure 9.7. Clutch size as a function of mass-independent basal rate of metabolism in flighted and flightless rails. (Modified from McNab & Ellis 2006.)

absence of eutherian predators? What about the occurrence of large flightless birds in Africa, South America, New Zealand, and Australia? Gondwana had no eutherian predators, which permitted the rheas, emu, cassowaries, and moas to evolve. Rheas had already attained a large mass before the great American interchange brought eutherian predators into South America. The presence of the ostrich in Africa is especially interesting. It evolved in Eurasia (Feduccia 1996) and entered Africa only about the Oligocene, when Africa approached Eurasia. Ostriches evolved in the presence of some early eutherian predators, but none of them were sufficiently cursorial to be effective predators on the ostrich's ancestors. The two families of eutherian carnivores that are effective cursorial predators, canids and felids, had not yet evolved (D. Steadman, personal communication). Obviously, large flightless birds, such as ostriches and rheas, passed a vulnerable stage with respect to the presence of eutherian predators, except for *Homo sapiens*, who makes all species vulnerable (see chap. 16).

The energy expenditures of New Zealand's endemic ducks, two of which are flightless, was also examined (McNab 2003a). These species had BMRs that were 70% of those of migratory anatids from the Northern Hemisphere, but somewhat surprisingly, flighted and flightless species from New Zealand did not have different mass-independent basal rates (fig. 9.8A). The fundamental correlation, then, appears to be with a sedentary lifestyle: flighted species endemic to New Zealand, such as the brown teal (*Anas aucklandica*

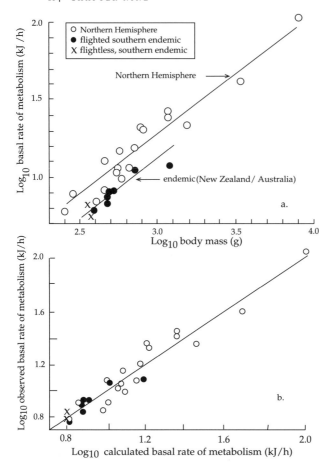

Figure 9.8. (A) Log_{10} basal rate of metabolism as a function of log_{10} body mass in Northern Hemisphere and flighted and flightless Southern Hemisphere anatids. (B) Measured basal rate of metabolism in anatids as a function of the basal rate of metabolism calculated from equation (9.2). (A modified from McNab 2003a.)

chlorotis), blue duck (*Hymenolaimus malacorhynchus*), New Zealand scaup (*Aythya novaeseelandiae*), and paradise shelduck (*Tadorna variegata*), as well as several anatids shared with Australia, are nearly as sedentary as the two flightless ducks, the Auckland Island (*A. a. aucklandica*) and Campbell Island teals (*A. a. nesiotis*).

These factors can be combined to estimate the basal rates of twenty-seven anatids. When body mass alone is used, it accounts for 87.6% of the variation in anatid BMRs because of the limited range in body mass (33:1), which leaves appreciable variation unaccounted for. The best estimate for anatid BMRs is given by the following equation:

$$\text{BMR (kJ/h)} = 147.0 \, (E) \, g^{0.784}, \tag{9.2}$$

where E is the dimensionless coefficient for location, which equals 1.43 for the Northern Hemisphere species and 1.00 for the New Zealand endemics.

This relationship now accounts for 93.6% of the variation in basal rates (fig. 9.8B), which potentially leaves room for other factors to influence anatid rates. What is most interesting here is that a flightless condition is not a significant factor because flightless species cannot be distinguished from volant New Zealand species that are nearly as sedentary as the flightless species (see fig. 9.8A).

Another group of flightless birds comprises the three kiwis (*Apteryx*) of the North and South Islands of New Zealand. These species have BMRs that equal 60% (brown, *A. australis*), 79% (great spotted, *A. haastii*), and 88% (little spotted, *A. owenii*) of the values expected from mass in birds (McNab 1996). They feed principally on soil invertebrates and face cool to cold temperatures. The kiwi with the lowest BMR, the brown kiwi, has the widest distribution and is found naturally at the lowest altitudes, whereas the other two species are (or were) found at altitudes up to 1,000 m (*A. owenii*) and 700–1,100 m (*A. haastii*) (Folch 1992).

Why do flightless birds have lower basal rates than flighted species? Flightless species share reduced pectoral muscle masses (McNab 1994a). For example, pectoral muscle masses are smaller in the flightless ducks than in the flighted ducks in New Zealand, and both groups of New Zealand ducks have smaller pectoral muscle masses than Northern Hemisphere species (fig. 9.9). A similar pattern occurs in flightless rails (Beauchamp 1989; McNab 1994a; McNab & Ellis 2006): pectoral muscle masses in flightless species are as small as 11.7% of total body mass, whereas they are as large as 14.5% in the flighted *Crex crex* (fig. 9.10). Notice that the rate of metabolism in herbivorous rails is greater at all pectoral masses and that the impact of flightlessness and a small pectoral mass is greatest in the omnivorous species. That is why the two largest herbivorous flightless rails have basal rates that appear to scale in a manner similar to those of flighted species in figure 9.5A. In other words, BMR in rails is a multiple function, at least, of body mass, food habits, and flight capacity (see eq. [9.1]).

The small muscle mass (1.7%) in *Atlantisia*, which is the smallest living flightless bird, may reflect the evolution of a flightless rail on a very small island (Inaccessible in the South Atlantic, 14 km^2) and the necessity to reduce its energy expenditure, which is accomplished by a small size (40 g), a low mass-independent basal rate (51%), and a very small pectoral muscle mass. (The smallest known flightless bird was the Stephens Island flightless wren [*Xenicus lyalli*]. Stephens Island sits between North and South Islands in New Zealand. The bird disappeared in 1894, apparently because of predation by the lighthouse keepers' cat.)

Other flightless birds also have small pectoral muscle masses. The kakapo (*Strigops habroptilus*), the largest living parrot, a flightless species endemic

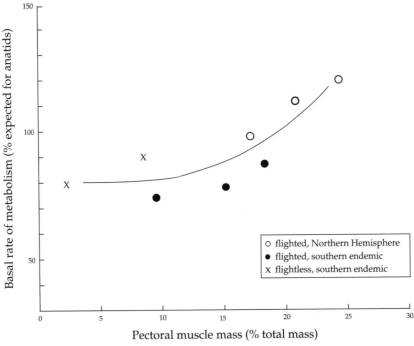

Figure 9.9. Mass-independent basal rate of metabolism as a function of pectoral muscle mass in Northern Hemisphere and flighted and flightless Southern Hemisphere anatids. (Modified from McNab 2003a.)

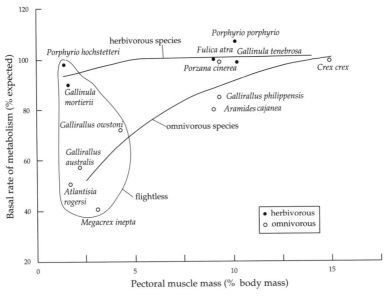

Figure 9.10. Mass-independent basal rate of metabolism as a function of pectoral muscle mass, a flighted or flightless condition, and food habits in rails. (Modified from McNab & Ellis 2006.)

to New Zealand, which has not had its BMR measured, has a pectoral muscle mass equal to 1.1%–1.5% of total body mass (Livezey 1992). It undoubtedly has a low BMR. And the brown kiwi has a pectoral muscle mass that is between 0.0% and 0.25% (McNab 1994a). A significant contributor to the reduced BMRs in all flightless species, therefore, appears to be a marked, often radical, reduction in the size of the pectoral muscle mass, but another contributor might be a reduction in heart mass. These data raise the possibility that the quantitative decrease in pectoral muscle mass reflects the time over which a species has been flightless: notice that, with the exception of *Atlantisia*, the smallest proportion of body mass that is pectoral muscle is found in the brown kiwi, which undoubtedly has been flightless since the early evolution of birds (see chap. 16).

Ectotherm prevalence

Given that a common response of intermediate to large endotherms to life on oceanic islands is a reduction of energy expenditure, it is not surprising that reptiles do well in such restrictive environments. For example, the "top" herbivores living on many tropical islands, including Aldabra, the Galápagos, and formerly the Mascarene Islands, Madagascar, Cuba, and Hispaniola, are, or have been, tortoises. Their low standard and field rates of metabolism permit a large mass (McNab 2002b). Large iguanids were found on the Galápagos, Fiji, and Caribbean islands.

The largest predators on some islands, such as New Guinea, the Marianas, the Carolines, the Marshalls, New Caledonia, and the Solomon Islands, were varanids, and terrestrial crocodiles were the top carnivores on New Caledonia (Buffetaut 1983; Balouet & Buffetaut 1987; Balouet 1991), Fiji (Worthy et al. 1999), and Vanuatu (Mead et al. 2002). The largest living lizard (possibly up to 250 kg [Auffenberg 1981], but usually no more that 150 kg), the ora (*Varanus komodoensis*), is the top predator on Komodo and adjacent islands, where it preys principally on deer, pigs, and occasionally people (Auffenberg 1981; box 9.2). The principal terrestrial predators of Australia were giant varanids, pythons, and crocodiles. The triumph of reptiles on islands depauperate of mammals is due to their low energy expenditures, which facilitate survival in environments that would force most large mammalian herbivores and carnivores to depress their energy expenditures, principally through a decrease in size, in an attempt to survive. In fact, the Komodo monitor has a field energy expenditure that is approximately 22% of the value expected for a mammal of equal mass (Green et al. 1991; McNab 2009c).

Evidence of the ability of ectotherms to flourish on a reduced resource base is found in Aldabran tortoises (*Dipsochelys dussumieri*), which may

BOX 9.2. ·

Living with a Dragon

Walter Auffenberg (curator of herpetology, Florida Museum of Natural History) lived on Komodo with his family for eight months, studying the Komodo monitor. He invited Wayne King (director of FMNH) and me to come for a month, but upon our arrival he warned us that when we got out of our tents in the morning, we should look for ora 34, which had become accustomed to human presence. (Weight was the basis for the numbering of oras; ora 34 weighed 33.5 kg. The number was written on the ora's side with white paint. The largest ora that was captured was number 54.) In his book on the monitor, Walter noted (p. 319) that "when 34 was released from the trap this afternoon, it walked straight toward both Wayne and Brian. Its mouth was slightly open, head held low and with roach slightly raised; the tail was arched and bowed. When it was approximately 2 m from both men, Wayne stepped out of the way. It veered only slightly from its original path, but it was not particularly intimidated by Brian's presence. It was only after it passed him that the lizard changed his slow stiff-legged walk to a more leisurely trot." After we left Komodo, a group of tourists came from Europe; three men reportedly in their twenties were walking down a pig or deer trail when a monitor came out of the brush, grabbed the middle man, and disemboweled him on the spot. Walter thought the attacker was number 34 because the attack occurred in its foraging area. I have often wondered whether our conditioning of this ora facilitated its attack.

· ·

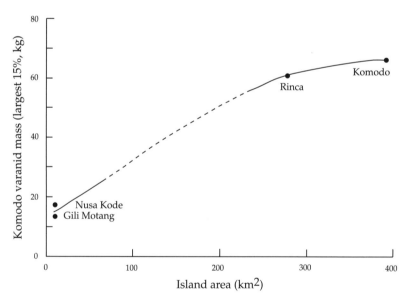

Figure 9.11. Body mass as a function of island area in the Komodo monitor, *Varanus komodoensis*. (Data from Jessop et al. 2006; McNab 2008c.)

maintain a community biomass of as much as 53,000 kg/km² (Bourn & Coe 1978). This biomass is about three times that of African mammalian herbivores. It occurs because tortoises have field energy expenditures that are about 3% of those of endothermic herbivores (McNab 2009c), which make an island existence no particular problem. The biomass density of the ora on Komodo, however, is only about 66 kg/km² (Auffenberg 1981), which reflects its higher mass-independent field rate, its top carnivore position, and the modest ungulate prey density of 2,486 kg/km² (Auffenberg 1981). That is, the low energy expenditures of ectotherms permit them to be effective specialists for the exploitation of low-energy opportunities (Shine 2005), as long as they are protected from eutherian predators. This ability of low-energy vertebrates to attain large masses and high densities on a limited resource base may also have applied to dinosaurs (McNab 2010). Yet even top ectothermic predators must adjust their energy expenditures on islands to reflect resource abundance: the mass of oras decreases strikingly with island size (fig. 9.11).

A striking discovery by Meike Köhler and Salvador Moyà-Solà (2009) showed that a Plio-Pleistocene dwarf (19 kg; Köhler 2010) bovid, *Myotragus balearicus*, from Majorca, had long bones characterized by a lamellar-zonal growth pattern, which is similar to that found in crocodilians. The long bones of this species "grew at slow and variable rates and ceased growth cyclically." This apparently led to an extended juvenile development period and life span, which probably reflected a (greatly?) reduced level of energy expenditure, or possibly a redirection of energy expenditure to reproduction and growth (Köhler 2010). Whether this type of development means that some physiological state intermediate between typical ectothermy and endothermy existed in this species is unclear; according to Köhler and Moyà-Solà (2009), "the zonal bone of *Myotragus* quite likely reflects seasonal fluctuations in metabolic rate and/or body temperature over an extended juvenile period in response to fluctuating resource conditions on the island." Furthermore, "in energy-poor environments where reptiles usually replace mammals, selection for energy saving may be so imperious that mammals may revert to some ectothermic-like state that includes both physiological and development plasticity." Like the flightless condition in birds endemic to islands, the characteristics of *Myotragus* undoubtedly hastened its extinction with the arrival of *Homo sapiens*.

CHAPTER TEN. *An Active Life*

One of the most important components of field energy expenditure is the expenditure associated with activity. Endotherms with high mass-independent basal rates of metabolism, including canids, felids, delphinids, seabirds, and shorebirds, generally have high levels of activity, whereas species with low basal rates, such as sloths, armadillos, anteaters, and flightless birds, have low levels of activity. These differences in activity may be associated with differences in body composition, as well as in food habits and behavior, which further magnify the differences in the energetics of active and inactive species. Unfortunately, only a limited number of measurements of the cost of activity have been made with the use of doubly labeled water (for a description of this technique, see box 11.1), which yields the most reliable data from free-living individuals under natural conditions.

Mammals

To investigate the cost of activity in small mammals, Corp et al. (1999) measured energy expenditure in wood mice (*Apodemus sylvaticus*) in the field. Their expenditures markedly increased with both the time active in the field and the distance moved (fig. 10.1). However, the energy budgets of mammals that show little activity, such as koalas and sloths, are probably little affected by the cost of activity. Unfortunately, no studies of the cost of activity in these mammals exist, but, as will be seen in the next chapter, the field energy expenditures of these species are exceedingly low, which implies low activity levels as well as low basal rates.

An important element of energy budgets in addition to the time and distance of movement is its intensity. Taylor et al. (1970, 1982) analyzed the minimal cost of running on a treadmill in mammals: the rate of metabolism increases with velocity of movement (fig. 10.2A) and body mass (fig. 10.2B). As an approximation,

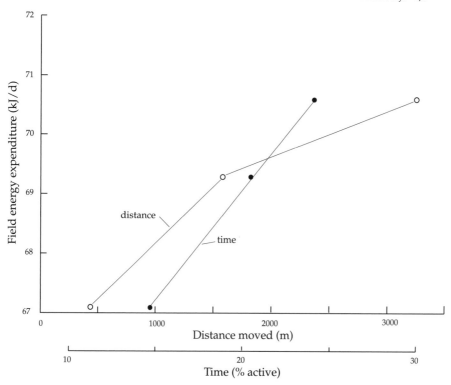

Figure 10.1. Field energy expenditure as a function of distance moved and time in wood mice (*Apodemus sylvaticus*). (Modified from Corp et al. 1999.)

$$M_{run} = a \cdot V \cdot m^{0.60} + c \cdot m^{0.72}, \tag{10.1}$$

where M_{run} is the cost of running (kJ/h), V is the velocity of running (km/h), the coefficient $a = 0.20$ kJ/km, and m is mass (g); the second term accounts for the increment of rate associated with body mass at $V = 0$, when c is about 1.4 times the coefficient for basal rate ($0.07 \times 1.4 = 0.10$ kJ/h). The cost of activity in some species is distinctive, such as the high cost of running in humans, whereas the horse conforms to the pattern seen in smaller species (see fig. 10.2B). (One can imagine how slow a treadmill, or treadbranch, would have to be to accommodate a sloth.)

Flight in mammals is not correlated with the basal rate ($P = 0.76$). In fact, the cost of flight in six species of small, nectarivorous glossophagine bats is 20%–25% less than its cost in small birds (Winter & Helversen 1998). Some of this difference may be associated with the smaller flight muscles of bats (Hartman 1963; Bullen & MacKenzie 2004). The difference in the flight of birds and bats has several consequences: (1) most microbats are restricted to

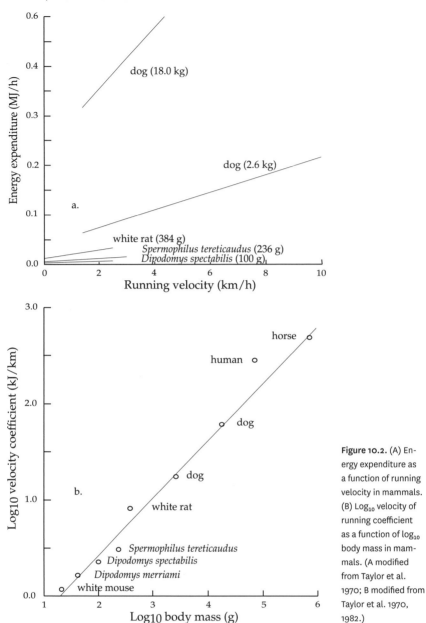

Figure 10.2. (A) Energy expenditure as a function of running velocity in mammals. (B) Log_{10} velocity of running coefficient as a function of log_{10} body mass in mammals. (A modified from Taylor et al. 1970; B modified from Taylor et al. 1970, 1982.)

continents or nearby island archipelagos, whereas the distribution of small birds is worldwide; (2) the bats most frequently found on oceanic islands in the South Pacific are megachiropterans, undoubtedly because of their large size and associated ability to fly long distances; (3) most temperate bats do not migrate to avoid harsh winter conditions, whereas most Arctic and many temperate birds do so; and (4) most temperate bats move to nearby shelters for hibernation in response to winter conditions. Therefore, mammalian flight, which opens some niches, also restricts ecological and behavioral diversity, as illustrated by the commitment of bats to a narrow set of nocturnal niches.

Birds

The principal cost of activity in most birds is the cost of flight. There are few direct measurements of flight cost. As one would expect, the cost of flight is highly variable. One factor influencing its cost is the type of flight. Flight in some species is simply the means of getting from one place to another, whereas in others it is the means by which they search for and capture prey. The cost of flight reflects the wing aspect ratio, which is defined as the square of the wingspan divided by the area of the wings, as well as other adjustments made to the flight apparatus to meet the requirements of the species' lifestyle. Species that spend more time in flight have higher aspect ratios (fig. 10.3) and lower costs of flight. The basal rates of 22 species of flightless birds averaged 74.0% $\pm$ 1.0% of the mean of 511 flighted species. Therefore, the cost of flight, as reflected in the basal rate of metabolism, is 35% greater energy expenditure than in flightless species.

The cost of flight is scaled to body mass (fig. 10.4). The bank (*Riparia*

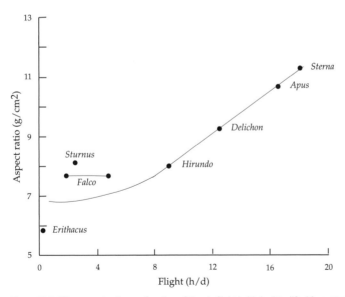

Figure 10.3. Wing aspect ratio as a function of time in flight in birds. (Modified from Masman & Klaassen 1987.)

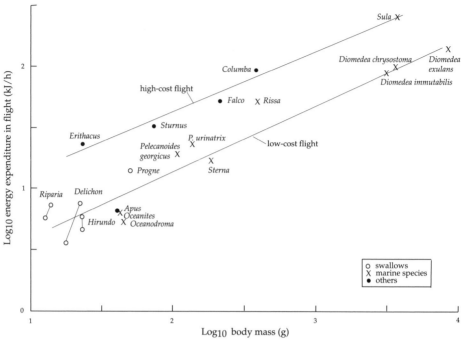

Figure 10.4. Log$_{10}$ energy expenditure in high- and low-cost flight as a function of log$_{10}$ body mass in birds. (Modified from Masman & Klaassen 1987.)

riparia) and barn (*Hirundo rustica*) swallows, house (*Delichon urbica*) and purple (*Progne subis*) martins, and the European swift (*Apus apus*), all of which capture their insect food in flight, have low energy expenditures per unit time in flight (Masman & Klaassen 1987), but as we shall see in chapter 11, they have high field energy expenditures because they spend much time foraging in flight. However, if their energy efficiency were less, their field energy expenditures would be even greater. These species have high aspect ratios, as does the sooty tern (*Sterna fuscata*). Other birds with a low cost of flight include species that use gliding flight, especially storm petrels (*Oceanites* and *Oceanodroma*) and albatrosses (*Diomedea*), which also have high aspect ratios. In contrast, species with low aspect ratios, such as the European kestrel (*Falco tinnunculus*), European robin (*Erithacus rubecula*), European starling (*Sturnus vulgaris*), and rock dove (*Columba livia*) have much higher costs of flight. The species with the highest cost of flight have expenditures that are about three times those of the species with the lowest cost, independent of body mass (see fig. 10.4). Other species are intermediate in the cost of flight, including diving petrels (*Pelecanoides*) and the black-legged kittiwake (*Rissa tridactyla*). As we saw in chapter 8, the cost of

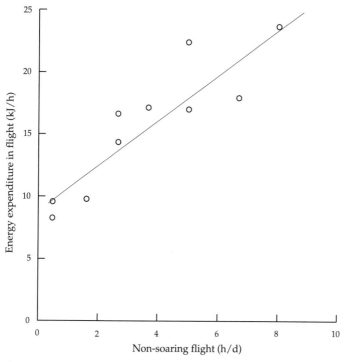

Figure 10.5. Energy expenditure in flight as a function of the time of non-soaring flight in the European kestrel (*Falco tinnunculus*). (Modified from Masman & Klaassen 1987.)

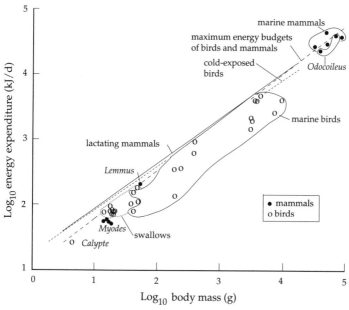

Figure 10.6. Log_{10} energy expenditure under various conditions as a function of log_{10} body mass in birds and mammals. (Modified from Kirkwood 1983.)

migratory flight in two species of thrushes was only 53% of the total cost of migrating 4,800 km (Wikelski et al. 2003b).

BMR might be associated with the persistent use of a gliding form of flight: do albatrosses, which glide extensively, have lower mass-independent basal rates than shearwaters, which tend to use more powered flight? The answer to this question is at best unclear: the mass-independent basal rates of three albatrosses equal 99%, 114%, and 142%, whereas those of five shearwaters equal 120%, 129%, 133%, 136%, and 147%. Or, does this variation simply reflect a relationship between flight form and body size?

The flight of predatory birds is highly variable. Some small species, such as the European and American kestrels (*F. sparverius*), often hover in pursuit of prey. At other times they soar or use powered forward flight. The energy expenditure during flight in the European kestrel varies with the time spent in non-soaring flight (fig. 10.5). Other predatory birds, such as *Falco* and *Accipiter*, use active pursuit, whereas buteos search for prey by extensive periods of soaring. Associated with these differences are variations in the size of the pectoral muscle mass and the heart (see chap. 3). Unfortunately, the costs of falconiform and vulture flight have not been measured.

Maximal energy expenditures

The maximal energy expenditures of birds and mammals are also scaled to body mass and proportional to $m^{0.72}$ (fig. 10.6; Kirkwood 1983). These expenditures were principally measured when the species were undergoing work, lactation, growth, or cold stress. Many of these values were obtained on domesticated species. What is especially interesting about the correlation of maximal energy expenditures with mass is that the maximal field expenditures approach the maximal laboratory expenditures, although field expenditures are highly variable, ranging from 10% to 80% of the maximal expenditures (Weiner 1992). Because of the approximately parallel scaling of the maximal expenditure to the basal rate, at least in mammals, Weiner (1989, 1992) suggested that the maximal rate is about 7 times the BMR, a view also proposed by Peterson et al. (1990). This proportionality gives the measurement of BMR even greater value.

CHAPTER ELEVEN. *Life in the Field*

Throughout much of this book, standard (i.e., basal) rates of metabolism have been the basis for an analysis of the adjustments of energy expenditures made by endotherms in relation to their behavior and to conditions in the environment. This is not to say that these rates are the most important, but they are relatively easy to measure, which accounts for the presence of data on almost 1,200 species of birds and mammals. A further important advantage of basal rates is that, if measured properly (see chap. 1), they reflect the same conditions and are therefore comparable in all endotherms. The rates that are presumably most important, however, are field energy expenditures (FEE): those that occur when individuals and species face life's real circumstances. But field energy expenditures are much more complicated to measure, which is why there are data on fewer than 200 endotherms.

Many people have measured the field energy expenditures of terrestrial vertebrates through the use of doubly labeled water (see box 11.1). The most extensive summaries of these data are by Nagy (1987, 2005) and Nagy et al. (1999). Data are available on 55 species of reptiles, 79 mammals, and 95 birds. These measurements are influenced by the species and by its activity level, food habits, behavior, and environment. Therefore, no one field rate will apply to an individual or species; rather, field rates will vary as these factors vary.

Field rates, like basal rates, are best described as power functions of body mass. Thus,

$$\text{FEE (kJ/d)} = 0.20 \ m^{0.889} \ (r^2 = 0.945) \text{ for reptiles,} \qquad (11.1)$$

$$\text{FEE (kJ/d)} = 4.82 \ m^{0.734} \ (r^2 = 0.950) \text{ for mammals,} \qquad (11.2)$$

$$\text{FEE (kJ/d)} = 10.50 \ m^{0.681} \ (r^2 = 0.938) \text{ for birds.} \qquad (11.3)$$

These equations can be compared with those that describe basal rates in birds (eq. [1.5]) and mammals (eq. [1.4]) because both pairs have similar powers, the principal difference are in the coefficients. The FEE curve of birds is (0.438/0.145) = 3.02 times their basal rate curve, and this ratio in mammals is (0.020/0.070) = 2.87. Therefore, field expenditures in birds and mammals are approximately three times those of basal rates. The correlation of FEE with BMR does not necessarily mean that they are functionally correlated because their correlation may simply be through their correlations with body mass. Whether these expenditures are functionally correlated will be explored by comparing their residual variations.

A comparison between FEE and the standard rates of reptiles is more complicated because the curves for standard rates depend on a fixed body temperature. Field expenditures of reptiles, however, reflect their body temperatures in the field, which vary with ambient temperature, the behaviorally selected body temperature, and possibly body size. These complications may lead to a higher power of the FEE curve (0.89) than found in standard curves (0.80), an interpretation dismissed by Nagy and collaborators (1999).

As we have seen in the case of basal rates of metabolism, field rates are greater in birds than in mammals and much greater in mammals than in reptiles (fig. 11.1): birds have FEEs that are about twice those of mammals, and mammals have FEEs that are about twenty times those of reptiles, subject to variations based on the difference in powers among these relationships. (Notice that the coefficient a is inversely correlated with the power b, as it is in eqs. [1.4] and [1.5].)

BOX 11.1. $\cdots\cdots\cdots\cdots\cdots\cdots\cdots\cdots\cdots\cdots\cdots\cdots\cdots\cdots\cdots\cdots\cdots\cdots\cdots\cdots\cdots\cdots\cdots$

Measuring Field Energy Expenditures

The principal means of *measuring* (as distinct from *estimating*) field energy expenditures is with water doubly labeled with deuterium and ^{18}O, both of which are rare in the environment. This method requires that an individual be captured twice, which greatly limits the amount of data accumulated. Upon capturing an individual for the first time, water doubly labeled as $D_2{}^{18}O$ is injected into the individual, the amount depending on the size of the individual and whether it is an ectotherm or endotherm. The animal is kept captive for a long enough time to allow the labeled water to become equally distributed throughout the body—from about 30 minutes in small species to a few hours in large species. Then a blood sample is taken, and the animal is released. The animal is (hopefully) recaptured after a significant, but not too long, period, usually a few days, when another blood sample is taken. The two blood samples are quantitatively compared to determine the amounts of deuterium and ^{18}O that have been lost. The loss of deuterium and ^{18}O via urine occurs in proportion to their abundance in body fluids. What is most important is the differential loss of ^{18}O via CO_2 and via H_2O. The decrease in deuterium accounts for the amount of ^{18}O lost as water. The remaining loss of ^{18}O reflects the loss of CO_2, which is a measure of the rate of metabolism over the given time period.

$\cdots\cdots\cdots\cdots\cdots\cdots\cdots\cdots\cdots\cdots\cdots\cdots\cdots\cdots\cdots\cdots\cdots\cdots\cdots\cdots\cdots\cdots\cdots$

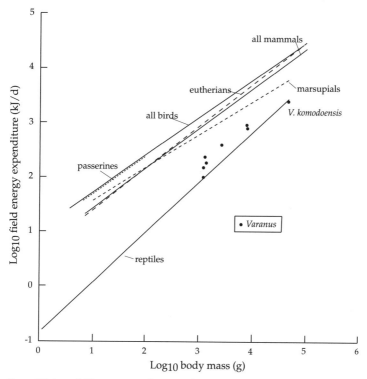

Figure 11.1. $\log_{10}$ field energy expenditure as a function of $\log_{10}$ body mass in birds, mammals, and reptiles, with additional data for monitors of the genus *Varanus*. (Modified from Nagy 1987 and Nagy et al. 1999.)

Reptiles

The reptile FEE curve has a high power, probably because small reptiles are very poikilothermic, reflecting the thermal environment in which they live. The four largest measured reptiles, which belong to the genus *Varanus*, including a 45.2 kg *V. komodoensis*, are active predators (McNab & Auffenberg 1976; Auffenberg 1981; McNab 2009c) (see box 9.2). Varanids usually have high field energy expenditures for reptiles (see fig. 11.1). For example, a 7.4 g lizard (*Podarcis lilfordi*) had a FEE equal to 1.5 kJ/d (Brown & Perez-Mellado 1994), whereas a 7.3 g pipistrelle (*Pipistrellus pipistrellus*), a temperate insectivorous bat, had a FEE equal to 29.3 kJ/d (Nagy 1994), 19.5 times that of the lizard, even though this bat goes into torpor. In contrast, the springbok (*Antidorcas marsupialis*), an African ungulate, at 43.3 kg had a FEE equal to 24.0 MJ/d (Nagy & Knight 1994), which is only 9.9 times the 2.43 MJ/d FEE of a 45.2 kg Komodo monitor (Green et al. 1991).

Birds

Fitted curves for field rates have some of the same limitations as those for basal rates. For example, the bird FEE curve has a lower power than the mammal curve, but five of the largest twelve bird species measured had low field rates, including three albatrosses, two penguins, and the ostrich, the last three being flightless. Much of the residual variation in field energy expenditures is related to differences in activity level, and as we have seen in chapters 3 and 9, flightless birds have "mammalian" basal rates. Passeriformes and Procellariiformes have high FEEs, whereas the Galliformes have low FEEs. These differences correlate with similar differences in BMR and with differences in activity level, as we will see shortly. Among birds, carnivores, nectarivores, and insectivores have higher FEEs than granivores. For example, high FEEs were found in swallows, reflecting their persistent pursuit of flying insects, and Anna's hummingbird (*Calypte anna*) had a high FEE, presumably because of the cost and time of hovering. A 39.1 kg mule deer (*Odocoileus hemionus*) had a FEE equal to 18.0 MJ/d, as did an 88.3 kg ostrich, which illustrates the lower power of the bird curve. Desert birds have lower FEEs than mesic species, a correlation that is compatible with the pattern seen in BMR.

The cost of activity can be high in marine birds. Energy expenditure is proportional to the time spent swimming in the Adélie penguin (*Pygoscelis adeliae*) (Chappell et al. 1993; fig. 11.2). Penguins leave their colonies to feed for extended periods. The king penguin (*Aptenodytes patagonicus*) forages for 5 to 10 days at an energy expenditure that is approximately 3.0 times the expenditure at the colony (fig. 11.3) and 4.6 times the "standard" rate. Feeding started at about 28 km from the colony (Kooyman et al. 1992).

Field energy expenditures in flying birds also increase with time spent foraging (fig. 11.4), as was seen in two marine species, the northern gannet (*Morus bassanus*) and southern giant petrel (*Macronectes giganteus*) (Birt-Friesen et al. 1989; Obst & Nagy 1992). The difference between these two species principally reflects the use of flapping flight by the gannet and the greater use of gliding flight by the petrel, although its gliding flight is much less efficient than that of albatrosses because of the petrel's higher wing loading (mass/wing area) (Obst & Nagy 1992) and the resulting necessity of mixing flapping with gliding flight. Albatrosses, which principally use low-cost gliding flight, had FEEs that varied from 53% to 84% of the values expected from equation (11.3). Albatrosses use flapping flight only 6%–7% of the time in flight (Pennycuick 1982), whereas the giant petrel uses flapping flight near 24% of the time (Obst & Nagy 1992). The highest mass-independent FEEs occur in some marine birds and mammals, which is not surprising, given their high BMRs and high activity levels.

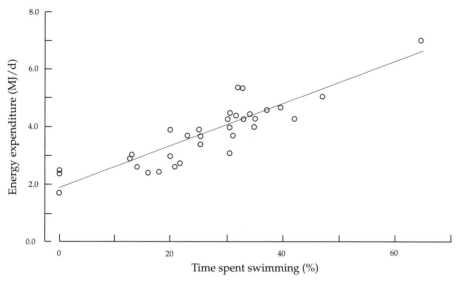

Figure 11.2. Energy expenditure as a function of time spent swimming in the Adélie penguin (*Pygoscelis adeliae*). (Modified from Chappell et al. 1993.)

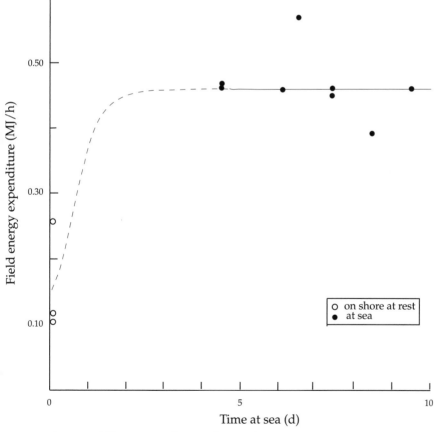

Figure 11.3. Field energy expenditure as a function of time on shore and at sea in the king penguin (*Aptenodytes patagonicus*). (Modified from Kooyman et al. 1992.)

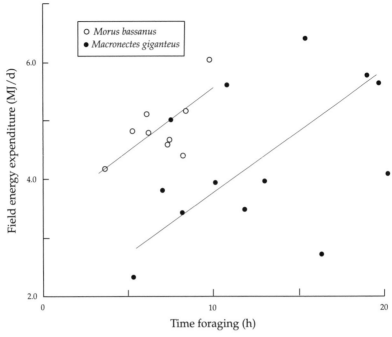

Figure 11.4. Field energy expenditure as a function of time spent foraging in two marine birds, the northern gannet (*Morus bassanus*) and the southern giant petrel (*Macronectes giganteus*). (Modified from Birt-Friesen et al. 1989 and Obst & Nagy 1992.)

Mammals

Factors other than body mass influence the field energy expenditures of mammals. They include phyletic affiliation, habitat, and food habits, but these factors influence both FEE and the power of the relationship, which makes an analysis difficult. Part of the problem is that any fitting of a power function depends on a large size range for the effect of size to be accurately estimated. As has been pointed out, if there is no range in mass, all of the birds or mammals of the same size will not have the same BMR or FEE as long as some ecological and behavioral diversity is present among these species.

Given these limitations, the principal phylogenetic difference in FEE found among mammals was between marsupials and eutherians (see fig. 11.1). Eutherians have a higher power (0.77) than marsupials (0.59):

$$\text{FEE (kJ/d)} = 10.1m^{0.59} \ (r^2 = 0.977) \text{ for marsupials,} \qquad (11.4)$$

$$\text{FEE (kJ/d)} = 4.21m^{0.77} \ (r^2 = 0.959) \text{ for eutherians.} \qquad (11.5)$$

As in basal rates, these estimated powers are not independent of the ecological characteristics of the species that constitute the sample if those characteristics occur unequally with respect to body mass. The difference between the infraclasses reflects the marine habits of the largest eutherians (which have high field rates that raise the fitted power of the eutherian curve) and the xeric distribution of the largest kangaroo (which has a rather low field rate, thereby depressing the fitted power of the marsupial curve). As a result of the ecologically narrow range of FEEs in marsupials, body mass accounts for a greater fraction of the variation in FEE in marsupials than in eutherians. No significant differences in FEE were found among the various mammalian orders, but many orders have had few species measured.

With reference to the influence of habitat, the power of the FEE curve was greater in desert species, which may reflect a significant difference in the ecology of large and small species. Many smaller species feed on insects or seeds and have low mass-independent FEEs associated with a propensity to enter torpor, which reduces their basal rate when normothermic, whereas the larger species, none of which enter torpor, are principally grazers or carnivores and have higher FEEs. Somewhat surprisingly, no correlation of FEEs with season was found, which may indicate that seasonal adjustments of insulation and microhabitat selection are sufficient to compensate for the seasonal change in conditions in desert environments.

Some species, because of distinctive aspects of their natural history, have unusual patterns of activity and energy expenditure. For example, during the breeding season, female Antarctic fur seals (*Arctocephalus gazella*) spend days with their pups and then go to sea to forage for food; like penguins, the time they spend at sea depends on the availability of food. At Bird Island, South Georgia, Antarctica, food, especially krill, was scarce in 1984 compared with 1985 (Costa et al. 1989). As a result, the seals' foraging trips in 1984 were twice as long as those in 1985 (fig. 11.5A). The principal factor influencing their field energy expenditures was the mass of the foraging female (fig. 11.5B). The mass gain by the foraging adult was about the same in both years, but in 1985 an adult required only about one-half the time to obtain the same amount of food as in 1984, which led to a mortality rate in fur seal pups in 1985 that was one-half what it was in 1984.

Some correlations of FEE with diet were found, but the limited sample size and diversity of the measured species limits our ability to draw clear conclusions. For example, the concept of "carnivores" is not simple: are we referring to the taxon or the diet? Nagy et al. (1999) concluded that "carnivores" (the taxon) have a higher scaling power than insectivores and herbivores. But not all of the "carnivores" are equally carnivorous (i.e., vertebrate-eating). The smallest species measured were *Bassariscus astutus*,

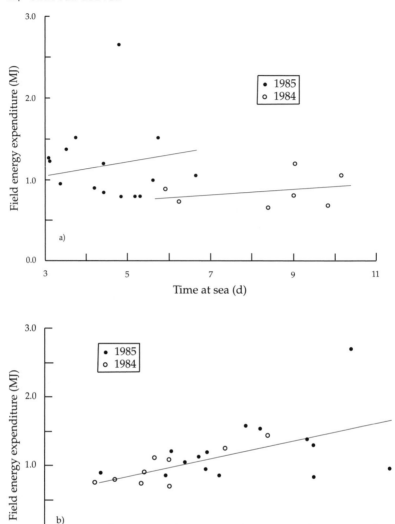

Figure 11.5. Field energy expenditure in Antarctic fur seals (*Arctocephalus gazella*) in 1984 and 1985 as a function of (A) time at sea and (B) body mass. (Modified from Costa et al. 1989.)

which feeds on insects, rodents, birds, and fruit and has a FEE that is 76% of the value expected from equation (11.2), and *Vulpes cana*, which lives in dry environments and feeds on small vertebrates and insects and has a FEE of 85%. The largest carnivores measured were the African hunting dog (*Lycaon pictus*, 187%) and the wolf (*Canis lupus*, 162%), which feed primarily on megavertebrates. The taxonomic definition of carnivory obviously is not trophically and energetically uniform, whereas a dietary definition would be more uniform. The size-dependent distribution of diets determines the high fitted power of body mass in relation to FEE in the Carnivora.

Speakman (2000) composed a comprehensive summary of FEEs in "small" mammals. Unlike the summary of Nagy et al., in which all the data were obtained using doubly labeled water, Speakman's summary included estimates from indirect calorimetry, basal and resting expenditures, time and energy budgets, and field heart rates as well as measurements obtained using doubly labeled water. The heterogeneity of these data leads to greater variation in the estimated FEEs, as well as uncertainty about the accuracy of estimates based on methods other than doubly labeled water. Speakman demonstrated, as is to be expected, that these FEEs correlate with body mass, but what is a bit of a problem is that his mean curve equals

$$\text{FEE (kJ/d)} = 6.88 \, m^{0.650} \, (r^2 = 0.888). \tag{11.6}$$

The data used by Speakman represented 72 species, with the two most deviant species, *Sorex araneus* and *Bradypus variegatus*, dropped. (As we have seen, *Sorex* shrews have very high BMRs and tree sloths have very low BMRs, so it is not surprising that their FEEs reflect these differences.) Still, the power of this curve is lower than that of equation (11.2) (0.734 vs. 0.650). Speakman's curve has a problem, however, for another reason: the coefficient 6.88 is 1.4 times that of Nagy et al.'s mammal curve. Some of this difference reflects Speakman's smaller mass range, 2,875/7.6 = 378.3-fold, whereas the range in the Nagy curve is 99,000/7.3 = 13,561.6-fold, 35.8 times greater than in Speakman's curve. The mass range influences the fitted power of the curve, and as the power decreases, the coefficient increases. The Speakman curve may operate for most "small" mammals, but it is unclear why the rufous rat kangaroo (*Aepyprymnus rufescens*), at 29 kg, or the three-toed sloth, at 4.2 kg, would be called small, although admittedly they cannot be confused with elephants. The fundamental difficulty with Speakman's curve is that it is a size-limited subset of the total mammalian relationship, which shortchanges the effect of body mass.

Factors other than mass that Speakman found to influence FEEs included ambient temperature, season, latitude, and food habits. A complication of the effect of season on FEE is that seasons are not independent of ambient temperatures, which are 10°C–20°C lower in winter than in summer for the species measured or estimated. However, when body mass and ambient temperature were included in the analysis, mammals in winter had FEEs that averaged 42% *lower* than in summer! That does not mean that FEEs are really lower in winter, but because the analysis was limited to "small" species, large species that cannot avoid cold ambient temperatures were not included in the study. Small mammals respond to winter by increasing insulation, being active in subnivean spaces, social huddling, and increased nest building, all of which reduce energy expenditure compared with what it would have been had they not made these adjustments, producing a pattern similar to that seen in Nagy et al.'s (1999) analysis. Furthermore, since most reproduction occurs in summer, the energetic costs of reproduction may have contributed to the higher summer FEEs. However, FEEs were positively correlated with latitude, even when mass and ambient temperatures were included. The reason for this pattern is unclear because of the correlation of ambient temperature with latitude. The increase in FEE with latitude may reflect an increase in plant production (Huston & Wolverton 2009).

FEEs are a significant correlate of food habits when mass, ambient temperature, and latitude are included in the analysis: high FEEs were found in species that fed on grass, insects, vertebrates, nectar, and fruit, whereas low FEEs were found in species that fed on the leaves of trees and seeds. Speakman found no correlation of FEE with taxonomic units when mass, ambient temperature, and latitude were included in the analysis. This finding seems appropriate because taxonomic units themselves are not independent of these factors; if they were eliminated from the analysis, FEE would strongly correlate with taxonomy.

Anderson and Jetz (2005) analyzed the factors influencing FEEs of endotherms. FEEs correlated with food habits and day length, but the only statistically different food habit, nectarivory, was associated with high rates. Birds had higher FEEs than mammals of the same size. However, FEE did not correlate with net primary production (NPP) when temperature and day length were included in the analysis. White et al. (2007a) came to a similar conclusion. NPP is difficult to separate from the influences of temperature and day length, however, because of its correlation with them, which presents a problem when one is attempting to determine the effective factors in a complicated relationship. Anderson and Jetz found that mass-independent FEEs (which were calculated as $FEE/m^{0.75}$) varied greatly at low latitudes, but were restricted to high values at high latitudes, which reflected their dependence on environmental temperature (i.e., FEEs were high at cold temperatures and showed a wide range at warm temperatures) (fig. 11.6). According to these authors, "low latitudes may allow a greater variety of feasible strategies than high latitudes," whereas the narrow range of FEEs at high latitudes represents "a fundamental physiological limit to the rate at which endotherms can process energy[, which] may limit the realized ecological niches and geographical distributions of endotherms."

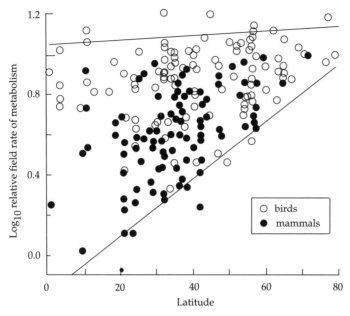

Figure 11.6. Field energy expenditure, relative to $m^{0.75}$, in birds and mammals as a function of latitude. (Modified from Anderson & Jetz 2005.)

Field and basal expenditures

The correlation of FEE with body mass, ambient temperature, latitude, and food habits and the correlation of BMR with body mass, food habits, thermal climate, habitat, a continental or island distribution, and a montane or lowland distribution (see chap. 3), raises a question: do FEEs correlate with BMRs? The absolute values of BMR and FEE, of course, must correlate at least through their mutual correlations with body mass; as we have seen, the ratio of FEE/BMR equals about 3:1. More important is the possibility that they correlate through their correlations with factors other than body mass. In other words, do the residual variations in these mass-independent energy expenditures correlate with each other? This question is important because of the extensive measurements of BMR that have been made and because of the aforementioned ecological analysis of the variation in BMR: is our commitment to measuring BMR justified, not only to determine BMR, but more importantly, to understand the performances of endotherms in the field?

This question has been occasionally asked and answered. Koteja (1991) raised this question with respect to thirty-one species of birds and twenty-seven mammals. He found that mass-independent FEE was strongly correlated with mass-independent BMR in eutherians (mainly rodents), but that the correlation was weak in birds and almost nonexistent in marsupials. He found no evidence of a proportionality between FEE and BMR, and he concluded that "the assumption that BMR is a reliable index of energy expenditure of free-living animals does not have a strong backing." Ricklefs et al. (1996), examining the same question from the viewpoint of phylogenetic contrasts, concluded that FEE correlated with BMR in mammals, but, using the discredited Sibley-Ahlquist cladogram, they failed to find a correlation in birds (see chap. 2). Speakman (2000) also explored the relationship between FEE and BMR. The residual mass-independent variations in FEE and in BMR are appreciable, especially in FEE, which, as noted, varies with ambient temperature, activity level, season, latitude, and diet. Speakman concluded that FEEs correlated with basal rates of metabolism in small mammals independently of body mass and "phylogeny" ($r^2 = 0.36$), but when two outliers, the common shrew (*Sorex araneus*) and the desert golden mole (*Eremitalpa namibensis*), were dropped, $r^2 = 0.12$, but remained statistically significant.

The consensus, therefore, appears to be that mass-independent FEE correlates with mass-independent BMR in eutherian mammals, although there is an appreciable amount of variation in this correlation and no simple ratio between these factors exists. In an attempt to clarify this situation, now that more data are available, measurements of FEE, derived only from doubly labeled water studies, on forty-one species of birds and forty-seven species of mammals that have had their basal rates measured have been assembled (McNab 2002a).

Birds

In birds, a positive correlation ($F = 34.97$, $P \leq 0.0001$) exists between the residual variation in FEE and the residual variation in BMR (fig. 11.7). The highest BMRs and associated high FEEs are found in marine species, including the alcids *Alle*, *Uria*, and *Cepphus*, the procellarid *Macronectes giganteus*, and the sulid *Morus bassanus*, all of which depend on high energy expenditures for flight. Many other marine birds whose BMRs have not been measured, and which are therefore not included in this analysis, have similarly high FEEs, including two auklets (*Aethia*, 168%; *Ptychoramphus*, 117%), two diving petrels (*Pelecanoides*, 181% and 186%), a prion (*Pachyptila*, 123%), and a tropicbird (*Phaethon*, 132%). Of interest here is the observation that penguins, which are flightless, yet live in cold water and are highly active, have intermediate BMRs and FEEs. The flightless ostrich (*Struthio*) has a low mass-independent BMR and FEE.

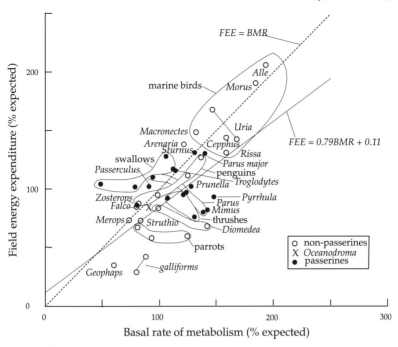

Figure 11.7. Mass-independent field energy expenditure as a function of mass-independent basal rate of metabolism in various birds. Two curves are indicated, one in which FEE = BMR and the other a curve fitted between FEE and BMR. (Data from Nagy et al. 1999.)

The correlation of mass-independent FEE with mass-independent BMR is described by the following equation:

$$\text{FEE} = 0.79\,\text{BMR} + 0.11. \qquad (11.7)$$

With $r^2 = 0.475$, much residual variation in FEE is independent of BMR, undoubtedly due to the influence of many factors, the most important of which is the cost of flight, as we have repeatedly seen. For example, the migratory, marine black-legged kittiwake (*Rissa tridactyla*) has a high BMR and FEE (see fig. 11.7). The migratory Arctic (*Sterna paradisaea*, 138%) and common (*S. hirundo*, 121%) terns have high FEEs, whereas (sedentary?) tropical terns, including the sooty tern (*S. fuscus*, 65%) and brown noddy (*Anous sozaidus*, 92%), have lower FEEs. None of these birds, unfortunately, have had their BMRs measured. The ruddy turnstone (*Arenaria interpres*), a long-distance migrant, has both a high BMR and a high FEE. Masman and Klaassen (1987) demonstrated that field energy expenditure in the European kestrel (*Falco tinnunculus*) correlates directly with the time spent in non-soaring flight (see fig. 10.5), although, given its mean activity period, this species has an intermediate to low BMR and FEE (see fig. 11.7). The rate of metabolism of the savannah sparrow (*Passerculus sandwichensis*) varies with the rate at which it visits its nest and with brood size (Williams 1987), but its mean FEE and BMR are intermediate (see fig. 11.7). Swallows have high FEEs due to their extended pursuit of insects in flight, although they have low to intermediate BMRs. (An exceedingly high FEE was measured in the tree swallow [*Tachyciena bicolor*, not shown in fig. 11.7], equal to 257% of the expected value in a species with a BMR equal to 115%.)

The importance of the cost of flight in determining FEE in birds is seen in two albatrosses (*Diomedea*), which have intermediate to high BMRs and intermediate to low FEEs (see fig. 11.7). These birds rely to a great extent on dynamic soaring, during which the cost of flight is only 2.5 to 2.7 times the expected basal rate (Costa & Prince 1987; Pettit et al. 1988); they use flapping flight only 6% to 7% of the time (see chap. 10). Field rates of metabolism in Leach's storm petrel (*Oceanodroma leucorhoa*) vary with the time spent at sea and whether the bird has an egg or chick (Ricklefs et al. 1986). But storm petrels are characterized by weak flight. Two species have FEEs equal to 83% (*O. leucorhoa*) and 88% (*Oceanites oceanicus*); another storm petrel (*Oceanodroma furcata*) has a BMR equal to 92%, which, *if* combined with the FEEs of the other two species, would put the FEEs of storm petrels between those of albatrosses and penguins (see fig. 11.7). There are several other groups of birds with low FEEs in association with low BMRs, including galliforms (*Callipepla* and *Alectoris*), some parrots (*Cacatua*, *Melopsittacus*, and *Neophema*), and a sedentary, terrestrial pigeon (*Geophaps*). Other galliforms have similar FEEs, including *Ammoperdix heyi* (40%) and *Centrocerus urophasianus* (71%), which spend little time flying, but have not had their basal rates measured, although they are undoubtedly low. Activity in birds, then, is an extremely important component of their field energy expenditures and accounts for much of the variation in their FEEs beyond that associated with body mass and BMR, but the amount of that variation depends on many external and internal factors.

Because a correlation of FEE with BMR, and apparently with activity level, exists in birds, a relationship should exist between FEE and BMR that incorporates activity level. In an attempt to find such a relationship, the following relationship should be explored:

$$FEE = a(F)BMR, \tag{11.8}$$

where F is the percentage of time in a day spent in flight, fully realizing the complication that all flight is not equal (see figs. 10.3, 10.4, and 10.5) and that for most species for which we have FEEs and BMRs, we do not have direct measurements of flight time. In fact, equation (11.8) should probably include another term, E, the efficiency of the cost of flight, a dimensionless fraction of the maximal cost of flight.

Mammals

Like birds, mammals have a pattern in the variation of FEE (fig. 11.8). The highest mass-independent FEEs and BMRs are found in arvicoline rodents (the highest FEE is found in the brown lemming [*Lemmus trimucronatus*, 220%], which is famous for its irruptive population numbers [see chap. 14]), terrestrial carnivores (*Canis lupus*, 162%), and marine mammals (*Zalophus californicus*, 227%; *Phoca vitulina*, 235%). The mass-independent BMRs measured in these four species are 132%, 128%, 205%, and 227%, respectively. Field energy expenditures have also been measured in some seals whose BMRs have not been measured, including *Callorhinus ursinus* (262%), *Arctocephalus gazella* (222%), and *Neophoca cinerea* (200%), which presumably would have high BMRs, as is generally the case in marine mammals and specifically in other seals (Ochoa-Acuña et al. 2009). What is fascinating here is that a tropical seal, the Galápagos sea lion (*Arctocephalus galapagoensis*), has a FEE that is only 44% of what is expected from body size! Either there is some problem with the measured FEE or the Galápagos sea lion has an unusually low BMR associated with its tropical (although rather cold-water) distribution.

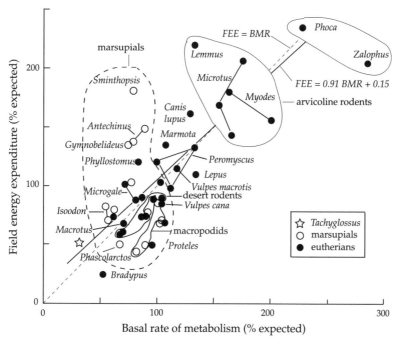

Figure 11.8. Mass-independent field energy expenditure as a function of mass-independent basal rate of metabolism in various mammals. Two curves are indicated, one in which FEE = BMR and the other a curve fitted between FEE and BMR. (Data from Nagy et al. 1999.)

The terrestrial, carnivorous African hunting dog (*Lycaon pictus*), which has a high FEE (187%), presumably has a high BMR: a reported thermoneutral measurement of about 225% (Taylor et al. 1971) may not have been postabsorptive. Obviously, we need more complete sets of data. The mammals with the highest mass-independent FEEs have the highest BMRs, which reflect the cost of continuous endothermy at a small mass, the foods they use, and their high levels of activity.

The mammals with the lowest FEEs, including the spiny anteater (*Tachyglossus aculeatus*, 53%), aardwolf (*Proteles cristatus*, 58%), koala (*Phascolarctos cinereus*, 50%), and three-toed sloth (*Bradypus variegatus*, 25%), have low BMRs (32%, 95%, 67%, and 52%, respectively). These animals have food habits that tend to depress BMR, including feeding on ants and termites or the leaves of trees, as well as low levels of activity. Other mammals with low FEEs include four macropods, which live in dry to desert conditions, and several seed-eating desert rodents.

An examination of the data in figure 11.8 reaffirms the conclusions of Koteja (1991) and McNab (2002a): no evidence of a correlation between residual variation in FEE and in BMR exists in marsupials, principally because there is little variation in marsupial BMR other than that produced by body mass, whereas a large variation is seen in marsupial FEE. For example, *Sminthopsis crassicaudata* had a very high FEE, as did two species of *Antechinus*, whose high FEE may reflect their intensive reproductive cycle, and *Gymnobelideus leadbeateri*, which is a very active species. Marsupial energetics is clearly organized in a radically different manner than eutherian energetics.

Unlike that of marsupials, the residual variation in the FEE of eutherians correlates ($F = 51.26$, $P < 0.0001$) with the residual variation in BMR (see fig. 11.8). Compared with that in birds, more of the variation in FEE is correlated ($r^2 = 0.531$) with the variation in BMR, and this correlation is almost proportional:

$$FEE = 0.91 \, BMR + 0.15. \tag{11.9}$$

This correlation and the similar one in birds suggest that correlations seen via body mass are actually functionally related. However, the functional bases for these correlations have not been identified. They may include body composition, as expressed through the effect of activity level on FEE, which, although incorporated into BMR, is more clearly expressed in FEEs, thereby contributing to the residual variation in FEE not accounted for by the residual variation in BMR.

What is clearly needed are many more *measurements* of the FEEs of a greater variety of birds and mammals, especially species from the tropics, only a small fraction of which have been measured. Given the comparative ease of measuring the BMR of species, further measurements of BMR should be encouraged in the attempt to understand the range of possible responses that species have, but these measurements do not completely substitute for measurements of field energy expenditures, especially in marsupials and to an extent in birds.

CHAPTER TWELVE. *The Limits to Geographic Distribution*

Climate undoubtedly greatly influences the geographic distributions of organisms (Wallace 1876). The concept of physiological limits to distribution in terrestrial organisms was summarized in an article written by C. Hart Merriam in 1894 entitled "Laws of temperature control of the geographic distribution of terrestrial animals and plants." Merriam suggested that the terrestrial world was divided into three principal climatic zones, boreal, austral, and tropical, each of which was subdivided with transitions between them. This article introduced the concept of life zones. In the Northern Hemisphere, species *"are restricted in northward distribution by the total quantity of heat during the season of growth and reproduction"* [Merriam's italics].

Because energy expenditure is so closely related to conditions in the environment, it might be expected to make some contribution to our understanding of the limits of geographic distribution of endotherms (Root 1988a). As reasonable as this may sound, the factors that limit distribution in most species are complex. For example, the limits may reflect a commitment to a geographically limited food supply, the presence or absence of other species, the abundance or absence of water, or the range of ambient temperatures as well as a limit to energy expenditures. The principal difficulty is separating the actions of these and other factors, which are usually intertwined.

The species that are most likely to show a thermal, or energetic, limit to geographic distribution are those that have made a radical physiological adjustment to a particular lifestyle in an extreme environment. Tropical rainforests, deserts, and polar environments often require such modifications, which may limit the ability of species endemic to these environments to move into adjacent environments. The question of the extent to which the physiological characteristics of species contribute to the limitation of their distribution remains. Unfortunately, few attempts have been made to answer it. What researchers often do is to correlate the limit of a species' distribution with a climatic isopleth. Such a correlation at best is suggestive of a possibility, but usually is insufficient to indicate what factors actually limit distribution. No general analysis of whether physiological characteristics limit the distributions of endotherms is available, so the examination of this possibility is most appropriate in some individual cases.

Latitudinal limits to distribution

Marsupials

Most marsupials have a tropical distribution, at least in the New World, but in Australia and Tasmania, many live in a mild temperate environment. Small marsupials in Tasmania often use daily torpor to survive cold conditions. In New Guinea, marsupials are found at altitudes up to 4,000 m (Flannery 1995): of 64 species, the maximal altitude of occurrence in 10 is less than 1,000 m; in 21, between 1,000 and 2,000 m; in 17, between 2,000 and 3,000 m; and in 16, above 3,000 m. Of course, New Guinea is tropical; the environment at 3,000 m is not cold. A few marsupials are found in southern South America (McNab 1982). They include *Thylamys elegans*, which uses daily torpor (Bozinovic et al. 2005), *Dromiciops australis*, which hibernates (Bozinovic et al. 2004), and *Rhyncholestes raphanurus* and *Lestodelphys halli*, both of which may hibernate. The few marsupials found at high altitudes in the Andes belong to the family Caenolestidae—namely, *Caenolestes* and *Lestoros*, which are found above 2,300 m.

A marsupial that penetrates far into a temperate environment in North America is the Virginia opossum (*Didelphis virginiana*), but it does so without entering torpor. It is found from South America and the Gulf Coast of North America northward to the Great Lakes, where it marginally penetrates southernmost Ontario, Canada. By introduction, it also occurs in southern British Columbia. Evidence suggests that *Didelphis* appeared in North America only in the Pleistocene (Hibbard et al. 1965); it has been found in post–Wisconsin glaciation archeological sites as far north as West Virginia and northern Ohio (Guilday 1958). Since then it has expanded its range, recently reaching its present northern limit.

Brocke (1970) and Kandu (2005) examined the northern limits to the distribution of this marsupial. Brocke analyzed its northern limit in Michigan. He noted that it correlated with the 70 freeze-day and 30 – 40 cm snow-depth isopleths, although the opossum was abundant only below these limits. In estimating these limits, Brocke relied on laboratory measurements of the responses of opossums to feeding and starvation. Survival in individuals increased with mass and with accumulation of fat stores in the fall. Males, which are the largest, can withstand the loss of 27% of their body mass over the 120 days of winter, whereas females can lose 44% and juveniles 29%. These losses correspond to a 44% decrease in lipid deposits in heavy individuals and a 72% decrease in light individuals. Because juveniles are the smallest individuals, this pattern reduces recruitment, an important factor in setting the northern limit to the opossum's distribution.

Kandu followed marked individuals over three years in Massachusetts. He demonstrated that the winters in this area were too cold to support a stable population, evidence for which was the limited survival of individuals, especially of juvenile females. For example, based on a model that was modified from that of Brocke, >67% of female juveniles must survive a winter to maintain a stable population, and this occurred in only 4 of the last 77 years in Massachusetts. The most important factor influencing the survival of individuals is their autumnal weight. This population survives principally near human settlements, where the animals depend on dumpsters, trash bins, and places where people leave food for wildlife.

Given the recent expansion of the opossum's range and the basis for the present limit, one can expect that the northern limit of its distribution will expand northward with global warming to occupy much of southern Canada.

Tree sloths

Species belonging to the two genera of tree sloths, *Bradypus* and *Choloepus*, are found from central Nicaragua south to northern Argentina. The northern geographic limit in Nicaragua is well south of the northern limits to the distributions of some foods used by sloths, including cecropia (*Cecropia* spp., Urticaceae), which is a common plant in disturbed areas in Guatemala, Belize, and southern Mexico. However, other trees whose leaves are preferentially eaten by sloths in Panama (Montgomery & Sunquist 1978), such as *Dipteryx panamensis* (Faboideae) by *Bradypus* and *Ficus trigonata* (Moraceae) by *Choloepus*, have northern limits to their distributions in Nicaragua, which suggests that the absence of preferred foods may contribute to limiting the distribution of sloths.

Yet energetics may also contribute to the geographic limits of tree sloth distributions. The apparent evolution of the two genera from different families of ground sloths (Webb 1985b) represents a remarkable convergence on a slow-moving, arboreal, folivorous lifestyle. A small muscle mass reflects this lifestyle: muscle constitutes about 24% of body mass in *Bradypus* and 27% in *Choloepus* (Grant 1978), whereas most mammals have muscle masses between 40% and 50% of body mass. Consequently, *Bradypus* has a mass-independent BMR of 52% and *Choloepus* of 55% (McNab 1978c). The resulting reduction in energy expenditure leads to highly variable body temperatures: in both sloths: they average 33°C–34°C at an ambient temperature of 20°C, 32°C at 10°C, and 29°C–30°C at 5°C. Low body temperatures in gut-fermenting herbivores may be unacceptable for extended periods because of the temperature dependency of fermentation and the adjustments made to facilitate handling of ingested plant secondary compounds (Janzen 1978).

The mean minimal January isotherm is about 24°C in lowland (23 m) Nicaragua, 22°C in Honduras, and 19°C in Guatemala, although in Flores, lowland Guatemala (115 m), it is 17°C. However, the mean minimal temperature may not be as important as the minimal temperature, which in lowland Guatemala may fall to near 10°C. These temperatures are likely to be unacceptable to sloths, especially if they last for any appreciable period.

Silky anteater

An interesting difference from the tree sloths is found in the silky anteater (*Cyclopes didactylus*). This species is the smallest arboreal mammal (250 g) committed to ant and termite eating. Like sloths, this species cannot control its body temperature at ambient temperatures <20°C. Its body temperature is 32°C at 20°C, 30°C at 15°C, and 23°C at 10°C, which reflects its small mass and a BMR equal to 62% (McNab 1984). It is found along the Caribbean coast as far north as the Isthmus of Tehuantepec, well beyond the northern limit of the sloths. It may face minimal temperatures lower than those faced by sloths, although these temperatures may be moderated by the proximity of the occupied area to the Straits of Tehuantepec. Why *Cyclopes* is tolerant of cooler temperatures is unclear, unless its tolerance reflects a less temperature-sensitive food source or digestive system.

Armadillos

Armadillos are distant relatives of sloths and tamanduas, all of which are characterized by low BMRs. Most armadillos are limited to the lowland Neotropics. However, the nine-banded armadillo (*Dasypus novemcinctus*) occurs in the United States. Its range in 1849 extended marginally across the Rio Grande in southern Texas (Humphrey 1974; Taulman & Robbins 1996) (fig. 12.1). Since then, its distribution has moved north and east. It was brought to Florida in 1924, after which its distribution expanded until it moved into southern Georgia. By 1980, the two disjunct distributions had merged, and the collective distribution moved north to southern Kansas and central Missouri. Humphrey (1974) suggested that two distributional requirements are annual rainfall >380 mm and <9 freeze days a year. Because rainfall has been decreasing along the Texas/New Mexico border, the

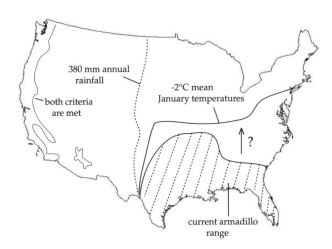

Figure 12.1. Limits to the current distribution of the nine-banded armadillo (*Dasypus novemcinctus*). Also indicated are two isopleths that may limit its distribution: 38 cm annual rainfall and a −2°C mean January temperature. The gap between the present distribution and the −2°C isotherm represents the area that may be occupied with time. (Modified from Taulman & Robbins 1996.)

Labels on map: 380 mm annual rainfall; −2°C mean January temperatures; both criteria are met; ?; current armadillo range

armadillo has withdrawn eastward (see fig. 12.1). The analysis of Taulman and Robbins (1996) agreed with the 380 mm rainfall limit and suggested a $<-2°C$ isotherm or <24 annual freeze days thermal limit. These conditions may set the ultimate northern limit on the distribution of this species, as long as the climate does not change. That limit would be defined by the Texas/New Mexico border to the west and to the north by central Kansas and Missouri, southern Illinois, Indiana, Ohio, and central New York (see fig. 12.1).

The limits to the distribution of this species undoubtedly have some physiological bases. The species has a BMR equal to 66% of the value expected from mass at a mass of 3.3 kg, although it is capable of maintaining a variable body temperature of 34.5°C down to an ambient temperature of 5°C (McNab 1980c), at least for short periods. Two additional factors that presumably influence the northern limits of its distribution are the armadillo's dependence on soil invertebrates as food, so that frozen, snow-covered ground would be unacceptable for any extended period (which relates to the 9 freeze-day limit), and the absence of a fur coat, which leads to a thermal conductance that is about twice that expected from mass. Water balance in the armadillo has not been studied, but is probably involved in the limit to its distribution into xeric environments. Of course, seasonal torpor is a possible adaptation to life in a cool- or cold-temperate environment, but no evidence has indicated that this species uses torpor.

Approximately thirteen species of armadillos reach their southern limits of distribution in Argentina (Redford & Eisenberg 1992); only five are found as far south as 40° S latitude (fig. 12.2). Of these, two species, *Zaedyus pichiy* and *Chaetophractus villosus*, are found in Patagonia as far south as the

Straits of Magellan. *Zaedyus* is small (1.7 kg) and has a low BMR (53%) and a highly variable body temperature (McNab 1980c); it may enter torpor on either a short- or long-term basis under winter conditions. *Chaetophractus villosus*, in contrast, is a much larger species (5.4 kg), with a basal rate equal to 54%, and is a good thermoregulator (McNab 1980c); its response to winter temperatures is unclear. Of interest is that *D. novemcinctus* does not penetrate as far into a temperate environment in Argentina as it does in North America (see fig. 12.2). The smallest armadillo, the pichiciego minor, or fairy armadillo (*Chlamyphorus truncatus*, 45 g), lives in central Argentina between 36° and 39° S. It has a dense fur coat under and at the sides of the carapace, but little is known of its biology and nothing of its physiology.

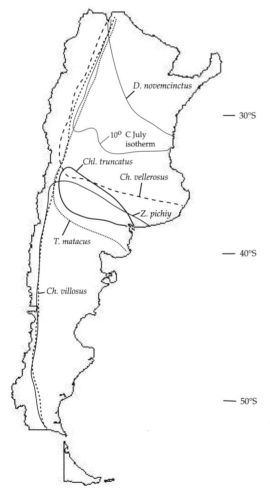

Figure 12.2. Southern limits to distribution in Argentina of six species of armadillos. Generic abbreviations: *Ch.*, *Chaetophractus*; *Chl.*, *Chlamyphorus*; *D.*, *Dasypus*; *T.*, *Tolypeutes*; *Z.*, *Zaedyus*. (Data from Redford & Eisenberg 1992.)

Marine mammals

Some marine mammals have distributions that appear to be limited by temperature. For example, the gray seal (*Halichoerus grypus*), which weighs 250–280 kg, gives birth in midwinter to pups that remain on land for 5 to 7 weeks, during which their mass triples. While the pups are on land, their lower limit of thermoneutrality is about −7.5°C. The northern limit of breeding sites in the North Atlantic is circumscribed by the January–February −7.5°C air isotherm (Nordøy & Blix 1985; Hansen & Lavigne 1997). This seal is thus excluded from breeding in Greenland, but breeds in Labrador, Iceland, and Scandinavia. Lower ambient temperatures during the period of pup fasting would probably be prohibitive (Hansen & Lavigne 1997).

Another North Atlantic seal, the harbor seal (*Phoca vitulina*), which weighs 80–220 kg, has a breeding distribution that is limited to cold waters. It gives birth in summer, but is limited in distribution to regions that have a July–August mean air isotherm ≤25°C. At higher temperatures, pups overheat. Because reproduction in this species occurs in summer, it breeds farther north than the gray seal, in locations including Greenland and Svålbard. The smaller harbor seal requires colder temperatures than the larger gray seal, reflecting the difference in breeding season. If these seals bred during the same season, one would expect that the larger species would preferentially breed in the colder climate.

A cetacean that probably is also limited to cold waters is the beluga, or white whale (*Delphinapterus leucas*). It weighs 700–1,200 kg and is found in polar waters of Alaska, Canada (including Hudson Bay), Greenland, and Russia. Some disjunct populations are found in the St. Lawrence River in Québec, in the Amur River, and off Sakhalin Island. These animals seem to be confined within the 10°C mean July air isotherm. Like other cetaceans, they would be expected to have high basal rates, although they are slow predators on fish and, to a lesser extent, on cephalopods and crustaceans. Measurements on three individuals by Kasting et al. (1989) indicated a rate of metabolism that averaged 202% of the value expected from mass at water temperatures between 12°C and 18°C, which were probably thermoneutral temperatures. Innes and Lavigne (1991) questioned the validity of this value, but they did so at a time when all high measurements of marine mammals were contested.

In contrast to the cold-water seals and cetaceans, the West Indian manatee (*Trichechus manatus*), which weighs 400–600 kg, requires a tropical distribution, so the focus of its North American distribution is peninsular Florida. In summer it may occasionally stray as far north as North Carolina, and there is even one record from Rhode Island. Manatees require water temperatures >21°C, so these northern strays must return to Florida before winter or find a warm shelter; if they do not, they will die. In captivity, manatees feed erratically at water temperatures between 18°C and 20°C and may refuse to eat at lower temperatures, although they may feed at lower water temperatures in the field, if they can find warm water that will permit higher body temperatures to aid digestion (Irvine 1983). Their lower limit of thermoneutrality is approximately 24°C (Irvine 1983), which is very high for such a large endotherm. This temperature may define long-term acceptable and unacceptable temperatures.

Manatees in Florida tend to congregate in the warm-water outflows from power plants and artesian springs in winter. However, that is not always adequate during a cold period. In January 2010, minimal water temperature in southern Florida was 7°C, and water temperature usually varied between 8°C and 11°C, a record low. At least 503 manatees were found dead in Florida in 2009–2010; the death count in normal years usually varies between 32 and 69. The sensitivity of manatees to cold water is associated with their very low basal rates; various estimates range from 20% of the value expected from mass for the West Indian manatee (Irvine 1983) to 50% for the Amazonian manatee (*T. inunguis*) (Gallivan & Best 1980). This temperature sensitivity is also characteristic of their relatives, dugongs.

One of the most remarkable geographic occurrences was that of Steller's sea cow (*Hydrodamalis stelleri*), a dugong that lived in subarctic waters of the North Pacific, where water temperature is about 5°C, and along the west coast of North America, where water temperatures are also low. Georg Wilhelm Steller, who was part of Vitus Bering's expedition to the Commander Islands, discovered this species in 1741; within 27 years, it had been hunted to extinction. The sea cow's tolerance for low temperatures undoubtedly related in part to its very large mass (ca. 4,000 kg, ten times the mass of the West Indian manatee). Obviously, we have no information on its energetics. Surely this species had a low BMR, but how low is unclear. This species was the last representative of a series of Pacific cold-water sea cows (Domning 1978).

Tropical bats

Some bats that are essentially tropical move marginally into warm-temperate environments, usually on a seasonal basis, but possibly year-round if adequate food (flying insects) is available. As we saw in chapter 8, this dichotomy in North America is clearest in species belonging to the family Molossidae. Several species move into the United States on a seasonal basis, including species of *Eumops* and *Tadarida*. *Tadarida brasiliensis* in Texas,

Oklahoma, Colorado, and New Mexico migrate to Central America in the fall to avoid cold winter temperatures and the absence of food, but those in Florida, where an adequate supply of flying insects is present throughout the year, do not migrate. The Molossidae is also the tropical family that penetrates the farthest into southern South America, where its southern limit is exceeded only by the Vespertilionidae, the insectivorous family that is widespread in all climates (fig. 12.3). Another approach to life in the temperate zone is hibernation, but it is not known in molossids.

What is it about the molossids that permits them to exploit warm-temperate environments, at least on a seasonal basis, other than being insectivorous? Could it be related to energetics? In terms of thermoregulation

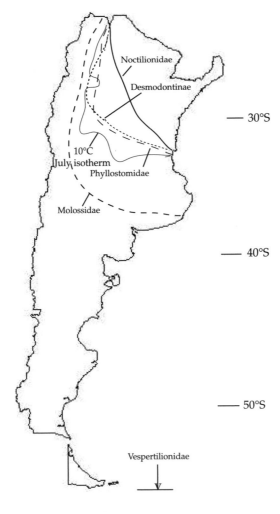

Figure 12.3. Southern limits to distribution in Argentina in four families and one subfamily of bats. (Data from Eisenberg & Redford 1992.)

and energy expenditure, only three species have been studied. These species thermoregulate better than is typical of vespertilionids, but with great variation, and they occasionally enter torpor (Leitner 1966; Licht & Leitner 1967; McNab 1969). They have a wing shape that suggests that they are fast long-distance fliers, which may facilitate their movement into and out of temperate environments.

Another family of tropical bats whose ranges approach temperate environments in both North and South America is the Desmodontidae, which is now considered a subfamily of the Phyllostomidae. It contains three species of vampires, all of which have low mass-independent basal rates (67%–88%) (McNab 1969, 2003b). Two of these vampires, *Desmodus rotundus* and *Diphylla ecaudata*, find their northern limit near the Texas/Mexico border and their southern limit in Uruguay, northern Argentina, and central Chile (see fig. 12.3). These limits roughly correspond to a minimal isotherm of 10°C during the coldest month in both North and South America.

The factors that might exclude vampires from a cool environment were examined in *Desmodus*, which is the most widespread species (McNab 1973), although the species that specializes on bird blood, *Diphylla*, has the highest BMR and may marginally enter southern Texas. Vampires have a foraging strategy that entails feeding once a night, usually within 2 hours of leaving the roost (Wimsatt 1969), which means that the blood meal must support the bat's energy expenditure for the next 24 hours. The blood meal is predominantly protein and water, which leads to resting rates of metabolism during digestion that are about 50% greater than basal rates. Weight loss is appreciable until the next feeding, principally because of urea and water excretion (McNab 1973) as well as high rates of evaporative water loss (McFarland & Wimsatt 1969). This water loss requires *Desmodus* to use humid roosts. Calculations indicate that life at cool to cold ambient temperatures, both at high latitudes and at high altitudes, would significantly raise the bat's energy expenditure (McNab 1973). The limit to meal size dictated by body mass and wing loading would prevent tolerance of ambient temperatures <10%. This limitation is probably greater for pregnant females. The occurrence of fossil *Desmodus* at localities in California and northern Florida suggests that the 10°C January isotherm was once farther north than it is today. With global warming, vampires would be expected to move northward into what is now the temperate zone. The third vampire, *Diaemus youngi*, is uncommonly found and little studied.

The sequence of penetration of temperate Argentina by bat groups is the same as that in the north-temperate zone: Vespertilionidae, Molossidae, Desmodontinae, and other Phyllostomidae (see fig. 12.3). Presumably this sequence conforms to a similar set of factors in both hemispheres. Those factors undoubtedly involve food habits (insectivorous bats penetrate the farthest, followed by vampires and frugivores). Other factors, or conse- quences, include the use of daily or seasonal torpor and possibly flight ca- pacity. All bats committed to continuous endothermy, in contrast to tem- perate species, have a tropical distribution.

Temperate bats

In striking contrast to tropical bats, almost all temperate species, which principally belong to the family Vespertilionidae, feed on flying insects, a seasonally unavailable food. Their reaction to conditions in a temperate climate is best seen at the interface of temperate and subtropical environ- ments. For example, three species of cave-dwelling vespertilionids are found in Florida (see chap. 8). One species, *Myotis austroriparius*, uses torpor dur- ing short cold spells in winter, but it becomes active during warm periods when insects are readily available. This species is found throughout Florida. The two other species, *Pipistrellus subflavus* and *M. grisescens*, are, unlike *M. austroriparius*, obligatory hibernators, which means that they are geo- graphically limited to regions in which sufficiently cold temperatures and shelters are available. The obligatory nature of their hibernation results from the shift from spring to fall copulation, with the result that females must store sperm until spring, which requires them to remain torpid.

The temperatures required by hibernating species depend on their body size and various aspects of their behavior: large species require lower ambi- ent temperatures for hibernation than small species, and clustering species (the cluster essentially produces a larger mass) require lower temperatures than solitary species (see chap. 8). Therefore, *M. grisescens*, which weighs about 10 g and is a clustering species, requires cave temperatures of about 5°C–10°C for hibernation. As a result, females of this species, which mar- ginally gets into northern Florida during summer, migrate northward to find winter cave temperatures cold enough for them to remain in torpor, because the coldest cave temperatures in Florida fall between 8°C and 13°C. A similar movement is found in some European cave bats, which migrate to altitudes between 700 and 900 m to attain cave temperatures between 3°C and 7°C for hibernation (Nagel & Nagel 1991). The pipistrelle, which weighs about 7 g, is a solitary hibernator, and requires cave temperatures no lower than about 15°C for an extended period of torpor, can hibernate as far south as central Florida, where cave temperatures are as low as 12°C to 15°C.

The use of torpor, hibernation, or obligatory hibernation by temperate bats, in association with their body size and various aspects of their behav- ior, influences the equatorial side of their distribution. Obviously, obligatory hibernators are limited to regions in which they can find sufficiently low ambient temperatures.

Naked mole-rat

The fossorial naked mole-rat from East Africa is highly sensitive to ambient temperature. It cannot maintain a body temperature above 30°C at ambient temperatures below 25°C (see fig. 1.9). The inability of *Heterocephalus* to withstand lower body and ambient temperatures prevents it from occupying altitudes above 1,000 m. However, as a result of its extremely low BMR and high thermal conductance, it can tolerate burrow temperatures from 30°C (see figs. 7.12 and 7.13) to 35°C, even at relative humidities >92% (see chap. 7). *Heterocephalus* "is a reliable 'indicator' species for arid conditions" (Kingdon 1974), which may well be a factor that permits, or requires, tolerance of high burrow temperatures because of low plant densities and a high solar heat input to the soil. This species is geographically limited by the Nairobi highlands to the west, the Ethiopian highlands to the north, and the Indian Ocean to the east.

Birds

In contrast to the situation in mammals, several generalized analyses of physiological limits to latitudinal distribution are available for birds. Root (1988a,b) performed one of the first such analyses. She found that approximately 60.2% of 113 North American passerines had northern range boundaries that corresponded to an average minimal January isotherm (Root 1988a). Root (1988b) then noted that the estimated maximal rates of metabolism at the northern limits of the birds' winter ranges were, like basal rates, a power function of body mass. She suggested that the northern limit to winter distribution in 14 species of passerines that have had their basal rates measured corresponded to an isotherm at which they had rates of metabolism equal to 2.45 times their basal rates of metabolism. Furthermore, this ratio at the northern limit of high-density populations in these 14 species was 2.13; that is, 87% of the value at the northern limit to winter distribution. An approximation of the ratio at the northern limit to winter distribution, estimated from allometric equations, was made for 36 species that have not had their BMRs measured; it equaled 2.64 times their estimated BMRs. Root concluded that "these findings strongly suggest that on a broad scale the winter ranges of a large number of passerines are limited by energy expenditures necessary to compensate for colder ambient temperatures."

Castro (1989), in an analysis of Root's argument, made several points.

(1) The calculation of the estimated metabolic limits to distribution in the 36 species without measurements of BMR consisted of circular reasoning, a view that has some merit because the calculation makes the assumption that all species conform to the standard metabolism-mass curve, which we have repeatedly seen is incorrect. (2) If the 2.5 ratio between the maximal and basal rates is uniform, "the only way birds can have a farther north [thermal limit to distribution] is by having a higher BMR (for any conductance)." (Of course, all conductances are not equal.) (3) The value of 2.5 is the product of chance. (4) If correct, the ratio 2.5 is simply the cost of maintenance at the northern limit, not a constraint. (However, if the 2.5 ratio is the product of chance in the 14 species in which BMR was measured, in which half of the values were 2.3 and 2.4, why should it be so uniform?)

In reply, Root (1989) argued that (1) the northern limit to distribution was not associated directly with BMR, but with the total rate of metabolism, which was a multiple of BMR. (Presumably, the total expenditure at the limit of distribution could vary with conductance and BMR.) Yet Root maintained that "species with the same body mass . . . implies that they will have *similar* BMRs." (That depends on what one means by "similar," which does not necessarily mean "the same.") She also argued that (2) random simulations of the limiting rate as a function of mass tell us nothing about the ratio. (What would be interesting would be to examine the 40% of species that did not have northern limits associated with an isotherm.)

Repasky (1991) also presented several challenges to Root's analysis. He stated that (1) relatively few—only 8% to 42%—of the passerines included by Root " actually" had northern limits that corresponded to isotherms. (2) Because total rates of energy expenditure increase with mass, he argued, shouldn't larger species have higher latitudinal limits to distribution? (3) Range limits do vary with body size: "birds of all sizes have range limits at the coldest temperatures studied, only a few large species have northern range limits in warm climes, whereas many small species do so." (4) "Species distributions are likely to be determined by the interactions of temperature and biotic factors [like food availability] rather than by simply a metabolic constraint."

Canterbury (2002) wrote another response to Root's analysis. He estimated the limiting environmental temperature for 28 species from allometric equations and measurements of BMR. He came to several conclusions: (1) "Estimates derived from physiological parameters predicted minimum temperatures at range boundaries more accurately than did allometric equations." (2) The distribution of energy expenditures at the geographic limits did not cluster at a thermal boundary. (3) The metabolic ratio at the geographic limit was smaller in insectivorous species than in those with other food habits. (4) The factors that influenced the distribution of species with other food habits were unclear.

These analyses touch on an important question: do the physiological characteristics of species influence their limits of distribution? The answer is undoubtedly yes, but it is a conditional yes. That physiology can uniquely define the limits to distribution is exceedingly unlikely because other factors, such as the availability of food, must be important as well (consider *Didelphis* in Massachusetts, or the observation that insectivorous passerines have lower metabolic ratios than species with other food habits, especially given the limited availability of insects in most cold climates). Behavior, too, must be important (consider *Desmodus* or *Myotis*). But the proposal by Root that the geographic limit of distribution in passerines is *influenced* by energy expenditure has played a valuable role, both by stimulating inquiry into this topic and by pointing out the importance of physiology to ecological phenomena. However, as is usually the case in biology, things are more complex than that.

Altitudinal limits to distribution

Many mammals and birds have a limited altitudinal distribution. The greatest diversity of endotherms is at low altitudes, which means that most species reach an altitude beyond which they are not found. This limitation may reflect the presence of other species with which the lowland species competes, the absence of resources used by the species, low ambient temperatures that the lowland species cannot tolerate, or some combination of these factors.

Unfortunately, few studies of energetics have examined the response of species to altitude. In the tropics, some species have made such extreme adjustments of energy expenditure that they may not be able to adjust to higher altitudes. This may well be the situation in tree sloths: the three-toed sloth (*Bradypus*) is found up to about 1,100 m, whereas the two-toed sloth (*Choloepus*) is found up to 1,800 m; this difference is possibly associated with a slightly higher BMR, larger mass, and potentially a different diet in *Choloepus* (McNab 1978c).

A group that may reflect a temperature-sensitive food supply is the Phalangeridae, the cuscuses. New Guinea has ten species, six of which have had their basal rates measured (McNab 2000b). Two of these six are found at high altitudes (1,400 to 3,900 m), one is widespread (0 to 2,700 m), and three are limited to low altitudes (0 to 1,500 m). Cuscus body size decreases with altitude, but their mass-independent BMRs are not correlated with altitude. However, when all marsupials are compared, montane species have lower BMRs than species limited to lowlands (McNab 2005b). The only apparent adjustment to high altitude is a thick fur coat in the mountain (*Phalanger carmelitae*) and silky (*P. sericeus*) cuscuses.

Sometimes the physiological patterns with reference to altitude are confusing. What might be expected was found in nectarivorous blossom bats (Pteropodidae) in New Guinea (Bonaccorso & McNab 1997). Two species, *Syconycteris australis* and *Macroglossus minimus*, have high basal rates (110% and 94%, respectively) and effective thermoregulation at altitudes between 600 and 2,100 m, but lower basal rates (67% and 57%) and a propensity to enter torpor at sea level. These species appear to have adjusted their thermal biology to permit activity at various ambient temperatures. In contrast, the frugivorous phyllostomids *Sturnia lilium* and *S. tildae* in lowland Brazil have high BMRs (123% and 131%) and a slight tendency toward a variable body temperature at cool ambient temperatures (McNab 1969, 2003b). *Sturnira erythromos* at 2,400 m in Venezuela (and 2,000 m in Ecuador) has a high BMR (157%), but a pronounced tendency to enter torpor (Soriano et al. 2002; McNab, personal observation). Why the highland *Sturnira* should go so readily into torpor in spite of its higher mass-independent basal rate is unclear, unless this reflects the availability of the fruits consumed by this species. Ambient temperatures up to 3,000 m in the tropics are not particularly low. In any case, this pattern is opposite to that found in the blossom bats in New Guinea.

Finally, the only extensive attempt to determine whether altitudinal distribution influences the energetics of birds has occurred in New Guinea (McNab, personal observation). A compilation of data on 78 species indicated that mass-independent basal rates correlated with altitude in a complicated manner. The maximal altitudinal distributions were grouped by 1,000 m intervals. If four intervals (0–1,000 m, 1,000–2,000 m, 2,000–3,000 m, >3,000 m) are combined with body mass, BMR is not a significant correlate of altitude, but it is when the first category is paired with the second through fourth categories. Then species belonging to categories 2–4 have a mean basal rate that is 1.31 times that of those belonging to category 1. That is, species limited in distribution to altitudes <1,000 m had lower basal rates than species whose upper altitudinal limit was >1,000 m, the causes of which presumably vary greatly from one species to another. This grouping of categories remained a significant factor when other factors were brought into the analysis.

······ PART IV. **Population Consequences**

CHAPTER THIRTEEN. *A Pouched*
(and Egg-Laying) Life

One of the goals of science is to find the general principles that control existence. Physics, of course, has been exceedingly successful in this quest, which has led other fields to seek similar goals. But the diversity of life has repeatedly demonstrated how living organisms have been able to exploit opportunities present in nature in such a manner that many of the "rules" controlling organisms must be modified by a series of conditional clauses (see box 1.2). This does not mean that life does not follow rules, but simply that most of those rules are complicated.

One place where such complexity appears is in the distinction between the energetics of monotremes, marsupials, and eutherians. As noted in chapters 3 and 5, eutherians that feed on ants and termites, or that are arboreal folivores or frugivores, have low basal rates of metabolism compared with the general mammalian curve, a condition that is also found in marsupials with similar habits. A monotreme anteater also has a low basal rate. Eutherians, marsupials, and monotremes with these food habits therefore appear to have similar mass-independent basal rates because they are dictated by factors external to the animals; that is, by the foods consumed (see chap. 5). However, eutherians that feed on grass or vertebrates have high basal rates, whereas marsupials with these food habits have either intermediate or low basal rates compared with the general mammalian curve. No living monotreme has these habits, and we do not know whether they ever did.

If the foods consumed influence BMR, why should some foods affect eutherians differently from marsupials? This appears to be a contradiction belying any generality. Or, if the difference in energetics between marsupials and eutherians is simply ascribed to the influence of "phylogeny" (i.e., history), as many contemporary biologists are wont to do, what does that mean? As normally practiced, that approach consists of circular reasoning: marsupials have low basal rates because they are marsupials, which are characterized by low basal rates, a statement that says nothing, except that marsupials have low basal rates. The real question is *why* marsupial carnivores and grazers have lower mass-independent basal rates than ecologically equivalent eutherians, since these foods *permit* high BMRs. Furthermore, we have just seen that the relationship of field energy expenditure to basal expenditure is radically different in marsupials than in eutherians.

The difference in energy expenditure between marsupials and eutherians actually reflects a fundamental inability of marsupials to exploit the abundance and quality of these food resources. To understand why this difference exists, we need to understand its basis beyond simply stating that it is due to "phylogeny," *which of course it is!* When mammals feed on ants and termites, they are usually *forced* by the food to have low basal rates (see chap. 5), whereas when they feed on vertebrates or grass, they are *permitted* by the food to have high basal rates, a possibility that can be exploited only by eutherians.

A difference in reproduction

The pivotal difference between marsupials and eutherians in their response to quality foods is based on a similarity and a difference in their means of reproduction. The similarity is that females of both groups implant young produced by sexual reproduction in their uteri. Sexual reproduction produces offspring that are genetically different from the mother. As a consequence, a pregnant female retains in her body one or more genetically foreign bodies in the form of offspring, which her body is programmed to reject. (This difficulty does not apply to monotremes and birds, which lay eggs.) For reproduction to be successful, immunological rejection must be countered without harming either the young or the mother, which is accomplished by isolating the offspring from the circulatory system of the mother. The fundamental difference between marsupials and eutherians is how they prevent the rejection of offspring.

Female marsupials protect themselves and their young by encasing the developing young in a shell membrane (Tyndale-Biscoe 1973), which "is a phylogenetic retention of a primitive amniote ontogenetic feature" (Lillegraven 1985). This membrane is similar to that found in monotremes and some oviparous reptiles, but differs from that in birds by the absence of calcification; it is unknown in eutherians (Lillegraven 1985). The shell membrane greatly reduces material and nutrient exchange between the mother's choriovitelline placenta and the offspring, thereby limiting the likelihood of rejecting the young, but also limiting the uterine development of marsupial young (Tyndale-Biscoe 1973). The membrane is shed during the last third of the gestation period, during which rapid organogenesis occurs in the offspring. Still, marsupials are born at an early stage of development compared with that of most neonatal eutherians. Much of marsupial development occurs after birth, when the mother transfers nutrients to the offspring via lactation and there is no threat of an immunological rejection of the offspring. Bandicoots, unlike other marsupials, develop a chorioallantoic placenta similar to that found in eutherians (Tyndale-Biscoe 1973; Padykula & Taylor 1976), but still do not produce young that are more developed at birth than other marsupials, probably because a trophoblast is not present. The development of neonatal marsupials is independent of the gestation period.

In contrast, eutherian females develop an elaborate chorioallantoic placenta from uterine tissue. The fertilized egg divides into a central cell mass (from which comes the embryo) and an outer sphere of cells, separated from the central cell mass by a cavity, the blastodermic vesicle. This outer sphere, called the trophoblast, is a structure unique to eutherians (Lillegraven 1985). It attaches to the placenta and is a "major barrier between the fetal antigens and the maternal antibodies . . . [partly due to a] noncellular external deposit" (Lillegraven 1976). The trophoblast has "physiological properties that allowed the unique combination of (1) intimate apposition of fetal and maternal tissues and circulatory systems (necessary to satisfy prolonged high embryonic energy demands); with (2) sustained, active morphogenesis (through which the fetus runs an ever increasing risk of rejection during gestation by the mother's immune system)" (Lillegraven 1985). The protection afforded by the trophoblast is so effective that some eutherians can sufficiently delay birth to produce precocial young, a behavior completely incompatible with a marsupial form of reproduction.

Consequences for marsupials

The difference in reproduction between marsupials and eutherians has remarkable physiological and ecological consequences. One is that no marsupial has a high basal rate of metabolism by general mammalian standards (see fig. 4.5). A high rate of metabolism occurs in eutherians largely because it facilitates high rates of reproduction by increasing the rate of exchange between a mother and her uterine offspring, thereby permitting high rates of uterine and post-uterine development and thus a reduction in generation time (McNab 1980b; see chap. 14). This reduction is achieved by shortening the gestation period, increasing the postnatal growth constant, and increasing the number of young produced by a female in a year (McNab 1980b). The observation that all eutherians noted for high population fluctuations, including arvicoline and some murid rodents, lagomorphs, and some predators, are characterized by high basal rates is compatible with the suggestion that a connection exists in eutherians between BMR and reproduction.

A high rate of metabolism cannot facilitate a high rate of uterine development in marsupials because of the presence of the shell membrane. All marsupials have average or low basal rates of metabolism (see fig. 4.5); the low rates are required by a highland distribution or an arboreal or burrowing lifestyle (McNab 2005b). The basal rate in 70 species of marsupials is given by

$$\text{BMR (kJ/h)} = 0.037(A \cdot S)\, g^{0.752}, \qquad (13.1)$$

where the dimensionless coefficients A and S stand for altitude and substrate, respectively. This relationship accounts for 99.2% of the variation in marsupial BMR. The coefficient A equals 1.25 in lowland species and 1.00 in montane species, and the coefficient S equals 0.66 in burrowing species and 1.00 in non-burrowing species. (The lowland coefficient >1.00 reflects reference to equation [13.1], the marsupial curve, not the all-mammal curve.) What is especially interesting is that the marsupial BMR is not correlated with food habits ($P = 0.27$) or the use of torpor ($P = 0.32$), nor is it correlated with an arboreal lifestyle ($P = 0.83$). As a result, eutherians have a much greater range in the residual variation of BMR when body mass alone is taken into consideration, whereas marsupials show a truncated distribution in mass-independent basal rate when compared with the all-mammal curve, at about 100% (fig. 13.1). As a consequence of the

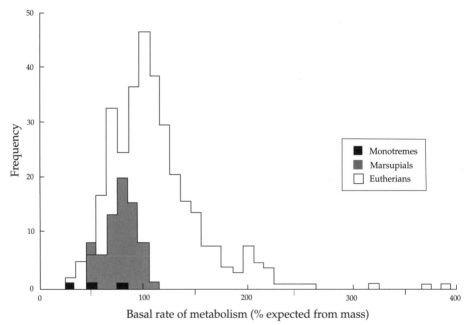

Figure 13.1. Frequency distribution of mass-independent basal rates of metabolism in monotremes, marsupials, and eutherians. (Modified from McNab 2002a.)

limited number of factors that impact BMR beyond the influence of mass in marsupials, mass accounts for more of the variation in marsupial BMR than was found in the general mammalian curve (eq. [3.1], $r^2 = 98.8\%$), which incorporates many more significant factors and possibly others that have been neglected.

An interesting aspect of equation (13.1) is that highland marsupials have basal rates that are 80% of those of lowland species, whereas highland eutherians have basal rates that average 114% of lowland species, and highland birds have basal rates that average 108% of lowland species (see chap. 3). The observation that marsupials have lower BMRs at higher altitudes is parallel to the general restriction of marsupials to warmer climates in the lowlands. Marsupial basal rates in burrowing species are 66% of those of non-burrowers. Therefore, the only factors other than mass that influence marsupial basal rates are those that *depress* basal rate: burrowing and life at high altitudes.

The constraint placed on marsupials by their form of reproduction makes their energy expenditures less sensitive to ecological factors. This observation led to the suggestion that the high BMRs of eutherians are principally associated with high rates of reproduction (see chap. 14), although, as we have seen, eutherians at small masses need appreciably higher basal rates to avoid entrance into torpor, an evasion denied all small marsupials.

Egg-laying mammals

And then there are the monotremes. Monotremes, like birds, lay eggs, but unlike those of birds, their eggs do not have a calcareous shell. The only monotremes that have survived are those that have food habits that are associated with very low-energy habit, which appear to give them a degree of protection from marsupials, the only mammals that might compete with them, given the distribution of living monotremes. Thus, ant- and termite-eating (*Tachyglossus*), worm-eating (*Zaglossus*), and aquatic worm-, cray-fish-, shrimp-, and fish-eating (*Ornithorhynchus*) monotremes have BMRs equal to 32%, 45%, and 76%, respectively, of the values expected from mass (see fig. 13.1): the higher basal rate of the platypus may reflect its food habits or its cool, aquatic habits. Whether monotremes ever successfully occupied high-energy niches is unclear, but undoubtedly they could never do so in the presence of eutherians, and possibly not even with marsupials. Monotremes probably survive today in Australia, Tasmania, and New Guinea because of the absence of eutherian equivalents (see chap. 14). A platypus, *Monotrematum sudamericanum* (Pascual et al. 1992, 2002), presumably another low-energy species, was found in the Early Paleocene of Argentina, reflecting a connection between South America and Australia, possibly through Antarctica. Clearly, monotremes were not able to survive later in the complex, competitive environment of South America. Monotremes were found in at least three fragments of Gondwana (Australia, New Guinea, and South America); were they ever in Africa or New Zealand?

CHAPTER FOURTEEN. *Energetics and the Population Biology of Endotherms*

One of the most expensive activities of organisms is reproduction. Its cost depends on many factors, including the number of young produced, their growth, development, and activity, and the period required for parental attendance, as well as weather and other external factors. Another factor potentially influencing the cost of reproduction is the means of reproduction. Does "internal" reproduction, as found in marsupials and eutherians, influence the energetics of reproduction in individuals or populations differently from an "external" means of reproduction, such as the laying of eggs by birds and monotremes? Or are all means equally responsive, or nonresponsive, to the influence of ecological factors?

These questions will be explored by examining whether some ecological factors affect the energetics of mammalian and avian reproduction and if so, their consequences for population biology. As seen in chapters 3 through 11, mass-independent basal rates correlate with many factors in the lives of birds and mammals. For example, a persistently limited quantity and quality of food, or a seasonal shortage of food, often *requires* a reduction in energy expenditure because of the necessity to balance the energy budget. On the other hand, some mammals and many birds have high mass-independent energy expenditures, which raises the question: why should any species *increase* mass-independent rates of metabolism when other species with similar habits and environments can get by with appreciably lower rates? And what are the consequences of ecologically similar species having different levels of energy expenditure? The usual explanation for high mass-independent rates of metabolism in endotherms is that they increase a tolerance of cold ambient temperatures or they increase mobility, reflecting a change in body composition. However, high energy expenditures depend on the exploitation of abundant, high-quality foods; otherwise they cannot be supported.

Although the availability of abundant and quality foods may *permit* an increase in the mass-independent BMRs and FEEs of their exploiters, their availability does not *require* an increase in a species' energy expenditure. For example, feeding on vertebrates permits high basal expenditures in felids (most fall between 114% and 165% of the values expected from the all-mammal curve [eq. 1.4]), canids (between 101% and 170%), and non-burrowing terrestrial mustelids (between 129% and 221%), not to mention otariids, phocids, and cetaceans, which have even higher energy expenditures. However, the larger dasyurids are all characterized by low BMRs, including *Dasyurus viverrinus* (86%), *D. geoffroii* (90%), and *D. maculatus* (77%), as well as the Tasmanian devil (*Sarcophilus harrisii*, 87%), which is mainly a carrion feeder. Whether the largest carnivorous marsupials have high or low field energy expenditures is unclear; small dasyurids *Sminthopsis* (15–20 g) and *Antechinus* (35–70 g) have high FEEs (138% to 181%; see fig. 11.8) even though they have low basal rates. Grazing ungulates have BMRs that vary between 117% and 227% and high FEEs (e.g., mule deer [159%] and springbok [198%]), whereas terrestrial marsupial grazers—macropods—have BMRs between 82% and 96% and FEEs between 45% and 86% (see fig. 11.8). What is responsible for the differential response of eutherians and marsupials to their food resources? It is inadequate to say that the difference is due to "phylogeny" and leave it at that. How is phylogeny involved with this difference in energetics?

All historical differences have causes and consequences. As stated in chapter 13, the basis for the difference in marsupial and eutherian energetics is a difference in the form of reproduction (Tyndale-Biscoe 1973; Lillegraven 1976; Lillegraven et al. 1987). As we shall see, high basal rates in eutherians have consequences for their population biology, but the absence of high basal rates in marsupials raises the question of what factors, other than body mass, if any, influence the population biology of marsupials.

The population energetics of eutherians

The growth of populations is described by

$$R_o = e^{rT}. \tag{14.1}$$

Therefore,

$$r = \ln R_o / T, \tag{14.2}$$

where r is the exponential population growth constant, R_o is the net rate of reproduction per generation, and T is generation time (Andrewartha & Birch 1954; Krebs 1972). When R_o is maximized and T is minimized, r is maximized; r_{max} represents the maximal capacity of a population to increase numerically by reproduction. R_o increases with litter size and with the number of litters that a female has during her lifetime. T decreases with a decrease in gestation period, the interlitter interval, and the period after birth required to reach the age of first reproduction. Variation in T is more effective at influencing r because R_0 is in the form of a logarithm (eq. 14.2). Body mass is the most important factor influencing the components of R_0 and T, and therefore r and r_{max}, but the rate at which energy is expended by eutherian mammals also has an appreciable impact (McNab 1980b).

The importance of body mass for the specification of the parameters of mammalian life history has been explored by Millar and Hickling (1991), whose paper opened a large literature on this topic. One statement made by these authors should be clarified. They noted "that interspecific patterns of life-history traits are more clearly related to body size than intraspecific patterns . . . , implying that the evolution of body size within species will not necessarily result in changes in life-history traits." The implication is correct, but the analysis is misleading. The reason why body size has a small intraspecific effect is because the intraspecific range in body size is very much smaller than the interspecific size range: no species has a hundredfold to thousandfold variation in body size. Therefore, intraspecific variation in life history traits is not accomplished by a change in body size. This leaves room for other factors to influence life history traits within a species.

As noted, R_0 is the product of litter size and litter frequency, as modified by the period over which a female is reproductive. Litter size is correlated with BMR (Glazier 1985a; Genoud 1988; Harvey et al. 1991; Stephenson & Racey 1995; Kalcounis-Rüppell 2007). Mean litter size in small (7–60 g) rodents increases with mass-independent BMR (fig. 14.1): litters are largest in arvicolines (*Microtus, Lemmus, Myodes*) and intermediate to small in neotomines (*Peromyscus, Podomys*) and heteromyids (*Chaetodipus, Liomys, Dipodomys*). This correlation is limited at the highest and lowest BMRs: litters in small mammals are not less than 2 to 3 offspring, even in species that have low BMRs, or greater than 6 to 8 offspring, even in species with the highest BMRs (see fig. 14.1). The correlation of litter size with rate of metabolism also reflects taxonomic affiliation because the level of energy expenditure is associated with food habits and environmental conditions and therefore with taxonomy. To untangle these connections is not easy.

In the small rodents examined above for litter size, the gestation period decreases with an increase in mass-independent BMR, reaching a low asymptote at about 20 days for BMRs between 150% and 250% of the value expected from mass (fig. 14.2), whereas for BMRs at 80%, the gestation period is about 30 days. Again, this pattern has a taxonomic component: arvicoline rodents have the shortest gestation periods, neotomines have intermediate periods, and heteromyids have the longest periods, although the arvicoline with the lowest basal rate, *Lemmiscus curtatus* (123%), has the longest gestation period (25 days) among arvicolines, similar to those found

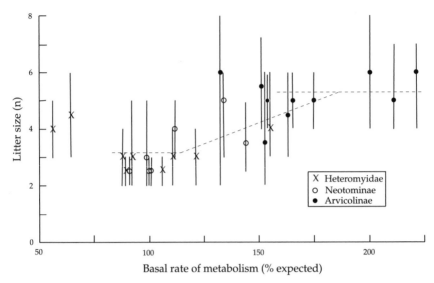

Figure 14.1. Litter size in rodents as a function of mass-independent basal rate of metabolism. (Data from Kalcounis-Rüppell 2007.)

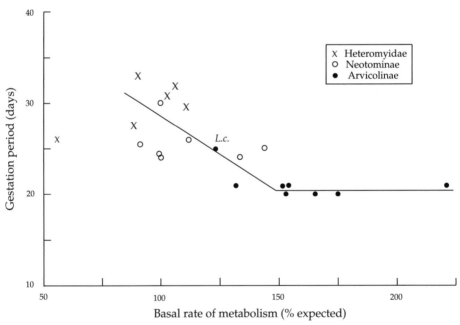

Figure 14.2. Gestation period in rodents as a function of mass-independent basal rate of metabolism. *Lc = Lemmiscus curtatus.* (Data from McNab 1980b and Kalcounis-Rüppell 2007.)

in neotomines. Among primates, two galagos that have BMRs equal to 86% and 90% have 124- and 136-day gestation periods, respectively, whereas the lorisid *Loris tardigradus*, of similar size, has a BMR equal to 61% and a gestation period of 167 days (Rasmussen & Izard 1988). That is, various groups have distinctive gestation periods, which are primarily determined by body mass and the developmental stage at which young are born, but are capable of being modified by variations in maternal metabolism. Symonds (1999) also found that gestation period decreased with an increase in (resting) rate of metabolism.

Another factor influencing reproductive output is the interlitter interval, the time that a female requires to reconstitute her reproductive tract after reproduction in preparation for the next period of reproduction. Interlitter interval decreases with an increase in mass-independent BMR (fig. 14.3). Again, arvicolines have higher BMRs and shorter intervals, neotomines are intermediate in both characters, and in the absence of data on heteromyids, murids with low BMRs generally have the longest interlitter intervals. The shortest interval in these rodents, at the highest BMRs, is about 20 days; the longest, at low BMRs, is 35 days.

The annual reproductive output of these rodents is the product of litter size and litter frequency (which varies inversely with interlitter interval). Because litter size in small rodents increases (see fig. 14.1), gestation period

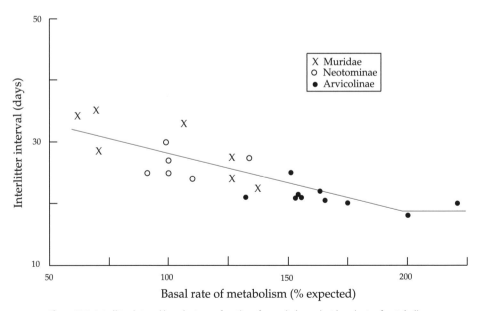

Figure 14.3. Interlitter interval in rodents as a function of mass-independent basal rate of metabolism. (Data from Kalcounis-Rüppell 2007.)

decreases (see fig. 14.2), and interlitter interval decreases (see fig. 14.3) with an increase in mass-independent BMR, litter frequency increases and annual reproductive output increases with mass-independent BMR (fig. 14.4), at least up to a flexible maximum of about 25 young per year, which undoubtedly depends on the availability of resources in the environment. These correlations are also with taxonomic affiliations, which are undercut by the overlap of BMR with the annual reproductive output in heteromyid and neotomine rodents (see fig. 14.4).

The total production of offspring by a female eutherian depends on two additional factors: (1) how long the female will live and (2) during what proportion of her life span she will be reproductive. Maximal life span increases with body mass and is inversely related to BMR (Hofman 1983). The maximal reproductive span is also proportional to body mass and inversely related to BMR (Calder 1984). Thus the level of energy expenditure fundamentally influences a species' reproductive performance, both on a day-to-day scale and over the long term.

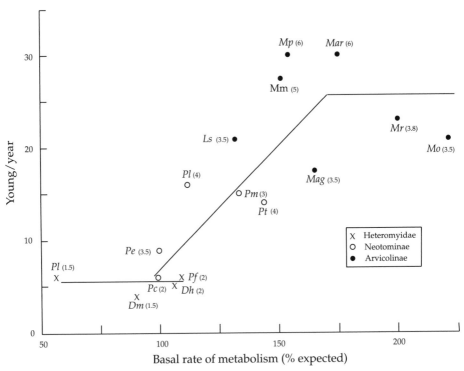

Figure 14.4. Young produced per year in rodents as a function of mass-independent basal rate of metabolism. Numbers in parentheses are the number of litters per year. (Data from McNab 1980b and Kalcounis-Rüppell 2007.)

The correlation of the parameters of reproduction with energy expenditure is correlated, in turn, with taxonomic affiliation—that is, phylogeny. So, what factor is most responsible for setting these parameters, other than body mass? Some hints can be found in the occasional species in a clade that has characteristics similar to species in other clades. For example, the sagebrush vole, *L. curtatus*, has, for an arvicoline, a low BMR and a long gestation period, both similar to those found in neotomines (see fig. 14.2). These character states probably reflect the vole's distribution in dry sagebrush flats in the Great Basin of North America, a distribution that is ecologically distinct from other arvicolines, which are usually tied to moist environments. As has been seen before, ecologically distinct members of a clade are often physiologically distinct as well, which suggests that their phenotypic characteristics depend on conditions in the environment and not on phylogeny, except to the extent that many clades are ecologically and physiologically uniform, and therefore limited. This conclusion raises another question: why are some clades more ecologically uniform than others? Rabosky (2009) addressed this problem. He concluded that diversity within a clade is not associated with clade age, but instead is ecologically limited.

Glazier (1985a) examined the relationship between energy expenditure and reproductive output in five species of *Peromyscus*. He demonstrated that energy expended for lactation, mass-independent postnatal growth rate, and litter size increased with mass-independent BMR. Similarly, Genoud and Vogel (1990) showed that peak energy intake during lactation in several species of shrews and rodents increased with mass-independent BMR (fig. 14.5), as it did in *Peromyscus* (Glazier 1985b). Shrews belonging to the genus *Sorex* have peak energy expenditures during lactation that are about three times their nonreproductive expenditures, whereas in *Crocidura* this increase is 1.2- to 1.7-fold. The higher expenditure in *Sorex* correlates with larger litter sizes and higher basal rates.

The time required to reach an adult mass varies with the individual exponential growth constant k:

$$m_2 = m_1\, e^{kt}, \tag{14.3}$$

where m_1 is mass at an earlier time, m_2 is mass at a later time, t is the time that has passed, and k is the growth constant. The principal factor influencing k is body size. In a sample of 28 species, McNab (1980b) described the following (slightly modified) relationship:

$$k = 0.190\, m^{-0.339}, \tag{14.4}$$

but this relationship did not account for all the variation in k ($r^2 = 0.777$). Much of the residual variation strongly correlated ($r^2 = 0.496$, $P \leq 0.0001$) with variation in mass-independent BMR: species with high BMRs grew

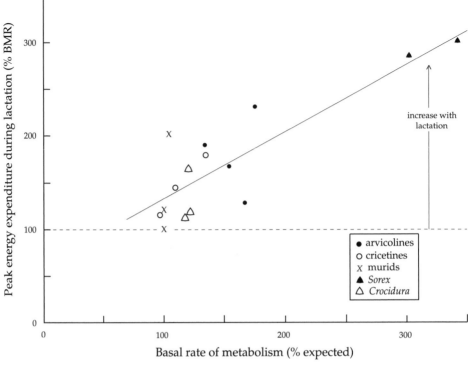

Figure 14.5. Peak energy expenditure during lactation in rodents and shrews as a function of mass-independent basal rate of metabolism. (Modified from Genoud & Vogel 1990.)

more rapidly than species with low BMRs (fig. 14.6). When mass and mass-independent basal rates are included in the analysis,

$$k = 0.165 \, (\text{BMR}) \, m^{-0.343}, \tag{14.5}$$

where BMR is the decimal fraction of the value expected from mass (eq. [1.4]); then $r^2 = 0.887$ (fig. 14.7).

Growth constants are high in arvicolines, intermediate in neotomines, and low in heteromyids, and the same pattern is found in their mass-independent basal rates. Furthermore, high growth constants and basal rates were found in a carnivore and a hare, intermediate to low growth constants and basal rates in fossorial rodents and domestic mammals, and even lower constants and rates in edentates, with the lowest in the African elephant. This correlation indicates that the time required for young to reach adult size decreases with an increase in mass-independent BMR. Symonds (1999) also suggested that growth rates are constrained by metabolism, especially in small species. Notice here and elsewhere that this form of analysis permits the precise prediction of the numerical properties of species, whereas a qualitative explanation does not account for the numerical characteristics of species. Readers need to decide which approach is the most informative.

Because R_o increases and T decreases with an increase in energy expenditure, r_{max} would be expected to increase with energy expenditure as well. Hennemann (1983) showed that the calculated maximal intrinsic rate of natural population increase varies with body mass in 46 species of mammals,

$$r_{max} = 4.91 \; m^{-0.262}, \tag{14.6}$$

but this relationship accounts for less than half of the variation in r_{max} ($r^2 = 0.476$). Hennemann argued that some of the remaining variation correlates with mass-independent BMR. When the residual variation in r_{max} is plotted

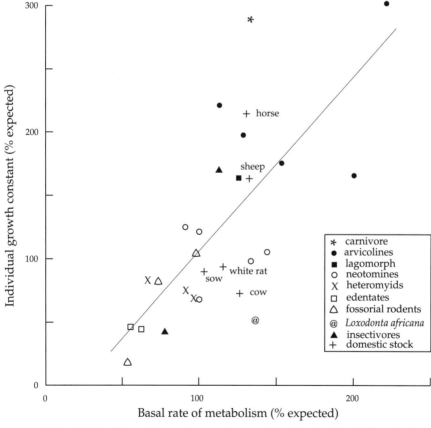

Figure 14.6. Mass-independent growth constant in mammals as a function of mass-independent basal rate of metabolism. (Modified from McNab 1980b.)

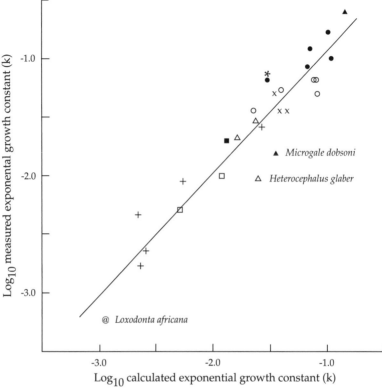

Figure 14.7. Log_{10} measured exponential growth constant as a function of log_{10} calculated exponential growth constant calculated from equation (14.5).

against the residual variation in BMR, species with a high BMR usually have a high r_{max} (fig. 14.8), but with great scatter. This pattern is seen when aquatic eutherians are not included in the analysis because their higher BMRs may reflect a thermally demanding environment, which does not necessarily translate into an increase in fecundity, as we saw in R_o. However, Schmitz and Lavigne (1984) argued that r_{max} in marine mammals is higher than in terrestrial species of the same size, possibly in association with their higher BMRs (Boily & Lavigne 1995, 1997; Costa & Williams 1999; Costa 2001; Williams et al. 2001; Boyd 2002; Williams & Worthy 2002; Ochoa-Acuña et al. 2009). The difficulty with the Schmitz / Lavigne analysis is that the correlation that they demonstrated was between r_{max} and mass-specific basal rates, which may occur simply because both factors are negatively correlated with body mass (McNab 1980b). This example shows why analyses must be mass-independent to eliminate passive correlations.

The variation in mass-independent BMR accounts for 25.5% of the residual variation in r_{max}, or 34.4% if aquatic species are excluded. Some of the variation in r_{max} independent of BMR conforms to a pattern (see fig. 14.8). For example, at BMRs greater than 100%, the r_{max}s greater than 200% are found in carnivores and lagomorphs of intermediate body mass, whereas the artiodactyls with similar BMRs have r_{max}s between 130% and 180%, and the few primates with rather high BMRs have low r_{max}s. So, obviously, factors other than body mass and energy expenditure influence r_{max}. The absurdly high r_{max}s calculated for *Didelphis virginiana* (448%) and *Ondatra zibethicus* (693%) are evidence of this. Clearly, we have much to learn about the factors that influence r_{max}.

If species with high mass-independent BMRs actually have higher R_os, lower Ts, and consequently higher r_{max}s, as a result of greater litter size and frequency, a shorter gestation period, and a shorter time required to attain adult size, then populations of eutherians with high BMRs would be expected to have greater amplitudes in abundance than species with low BMRs. Demonstrating this phenomenon might be difficult, however,

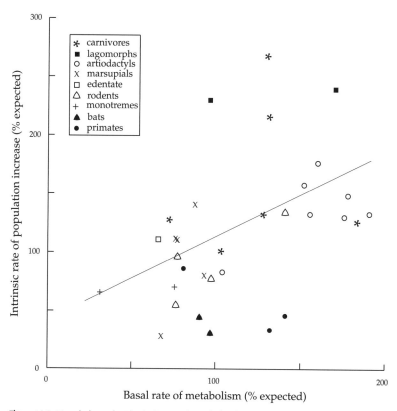

Figure 14.8. Mass-independent intrinsic rate of population increase as a function of mass-independent basal rate of metabolism. (Modified from McNab 1980b.)

because population fluctuations in nature reflect the environmental conditions encountered as well as the characteristics of the species. The conditions faced by a population in one place and time may not be equivalent to those faced by other populations of the same species at another place and time.

The best way of getting around this difficulty would be to examine comparative studies of species in the same environment at the same time, so that all populations in a study would be facing similar physical conditions. When such comparisons were made (McNab 1980b), the amplitude of population fluctuations in an area increased with the mass-independent BMR of the species (fig. 14.9). Thus, within a habitat, arvicoline rodents had higher population amplitudes than neotomines, which usually had higher amplitudes than heteromyids. Indeed, that is why some arvicoline rodents, such as *Lemmus trimucronatus*, *Dicrostonyx groenlandicus*, and *Microtus oeconomus*, and lagomorphs, especially *Lepus americanus*, are famous for large

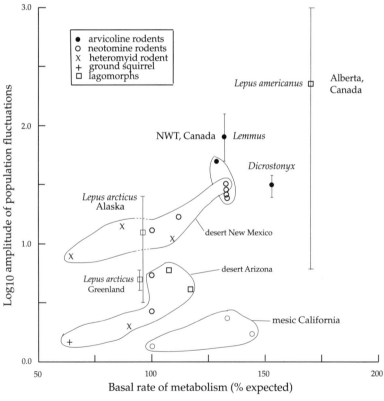

Figure 14.9. Log$_{10}$ amplitude of population fluctuations as a function of mass-independent basal rate of metabolism in mammals found in the same environments. (Modified from McNab 1980b.)

population fluctuations, which result from large litter sizes and high litter frequencies. This extreme ability to take advantage of hospitable conditions rapidly occurs in species with the highest basal rates, which directly points to the association of r_{max} with mass-independent energy expenditures as represented by BMR and its correlation with FEE (see chap. 11).

In a follow-up to this analysis, Kurta and Ferkin (1991) examined population fluctuations in two closely related arvicolines, *Microtus pennsylvanicus*, a species prone to high, multi-annual population fluctuations on the North American mainland, and *M. breweri*, a species endemic to sand dunes on Muskeget Island, Massachusetts, which shows only minor population fluctuations (Tamarin 1977). *Microtus pennsylvanicus* had a higher proportion of females that were pregnant in summer (0.60 vs. 0.46), and in winter (0.10 vs. 0.00), a larger litter size (4.5 vs. 3.4), and a 30% smaller mass at sexual maturity than *M. breweri*, although *M. breweri* had a higher growth rate (Tamarin 1977). As expected, *M. pennsylvanicus* had a higher mass-independent BMR, which equaled 154% of the value expected from mass, than *M. breweri*, whose BMR was 121% (Kurta & Ferkin 1991). Kurta and Ferkin recommended that the potential correlation of population demography with energy expenditure be further tested by examining closely matched species pairs, a good suggestion that has been rarely followed (see fig. 14.9).

A potential impact of these differences in population ecology is that species that have a higher mass-independent BMR and a higher r_{max} would be expected to exclude species using the same resources that have a lower mass-independent BMR and a lower r_{max}. An informative, if inadvertent, experiment to test this idea resulted from the human introduction of snowshoe hares (*L. americanus*) to Newfoundland, where the Arctic hare (*L. arcticus*) was endemic. Snowshoes spread throughout the island, excluding Arctic hares from the forests, even though the snowshoe hare is only one-half (1.6 kg) the size of the Arctic hare (3.0 kg), which is now confined to the barren grounds (Cameron 1958). This exclusion reflects the higher mass-independent basal rate (170% vs. 96%), a greater range in population fluctuations (see fig. 14.9), and a presumptively higher reproductive output in the snowshoe hare.

Grant (1972) summarized a series of studies on competitive interactions in pairs of rodents in which both species had had their BMRs measured. He described eleven cases in which one species excluded another. However, in only four of those species pairs did both species belong to the same genus (*Peromyscus*, *Microtus*, and *Tamias*). Intrageneric comparisons increase the likelihood that these pairs would be in competition should they be found together in the field. In each of these four cases, the species with the highest mass-independent BMR won the competition (table 14.1). Intergeneric

TABLE 14.1. Competition and energy expenditure

Winners (% BMR)	Losers (%BMR)
Microtus arvicolis (175)	Microtus agrestis (165)
Microtus oeconomus (221)	Microtus agrestis (165)
Peromyscus maniculatus (133)	Peromyscus oreas (123)
Tamias amoenus (150)	Tamias minimus (138)

pairings, however, gave inconsistent results, which may indicate that factors other than BMR and reproduction were responsible for the results. For example, *M. pennsylvanicus* (BMR = 154%) won over *P. leucopus* (BMR = 112%), but these two genera often coexist in the field, in part because their food habits and microenvironments are different. However, if the level of energy expenditure in two congeneric species is similar, other factors may determine the competitive outcome. For example, *Microtus pennsylvanicus* has a BMR equal to 154% and *M. montanus* a BMR equal to 151%—no difference—so it is not surprising that *M. montanus* excludes *M. pennsylvanicus* in dry conditions and the reverse occurs in wet environments (Grant 1972).

The proposal that energetics influences the population ecology of eutherians has been criticized. The principal complaints have been fourfold: (1) that the statistical bases for the proposed correlations are faulty (Hayssen 1984); (2) that there is no *theoretical* reason why a population's maximal rate of increase should correlate with the *basal* rate of metabolism (Hayssen 1984); (3) that data obtained on white mice do not show a correlation between the parameters of life history and energetics (Hayes et al. 1992); and, of course, (4) that most variation in life histories is related to phylogeny (Harvey et al. 1991; Lovegrove 2009).

Lovegrove (2009), using independent contrasts, made the argument that "since age at first reproduction is directly influenced by an individual's growth rate, . . . the age was not partially correlated with either body mass, growth rate, or BMR. Similarly, growth rate was not correlated with BMR." This view depends on the use of a phylogenetic "correction." However, the residual variation in the growth rate, as represented by the residual variation in the growth constant, is correlated with the residual variation in BMR, which in combination with body mass now accounts for 88.7% of the variation in k (see fig. 14.7). The principal reason that the use of independent contrasts does not find these correlations is that this technique ignores the correlation of character states with phylogeny.

These criticisms should be addressed. (1) It is clear that determining the factors influencing r_{max} is exceedingly complicated, which makes any statistical analysis suspect; such analyses can always be improved. A persistent procedural requirement is that the influence of body mass on a species' character states be removed because so many of the characteristics of species are correlated with mass. These individual correlations may lead to an apparent correlation between those characteristics because of their mutual correlation with mass—that is, a correlation that is not functionally based. The analysis used here has attempted to bypass these inappropriate correlations. (2) With reference to the absence of a theoretical justification for an influence of BMR, the original article (McNab 1980b) was based on observations, not theory. Most theories of complicated phenomena are simplifications that can be improved with the *acquisition of data*, an approach that distinguishes science from most other intellectual inquiries. Besides, reproduction obviously requires appreciable energy expenditures, so it is appropriate to ask what determines those expenditures and what consequences they have. Examination of the correlation of populational parameters with BMR is convenient because so many comparable measurements of BMR are available, and it can be viewed as the first step in identifying the theoretical basis for a species' performance. (3) Whether an intense examination of one strain of white mouse tells us anything about natural systems is open to question because the great diversity in behavior, food habits, and environment found in the native fauna does not apply to this degenerate organism, which could not survive if released into the wild. (4) The tendency to attribute all phenotypic differences among species to phylogeny completely ignores the fact that a large variety of characteristics are correlated with phylogeny, which makes the determination of the unique influence of phylogeny difficult to isolate from the influence of associated factors (see chap. 2). Phylogenetic analyses give interpretive preference to the impact of phylogeny. Even if a particular pattern in a species' characteristics is associated with phylogeny, as we have seen repeatedly in the comparison of arvicoline, neotomine, and heteromyid rodents, as well as with marsupials, the factors responsible for the association should be identified, as was done with the difference in BMR between marsupials and eutherians. Any analysis of the occurrence of character states that simply ascribes it to "phylogeny" and does not explore the basis for this association is superficial. The value of the approach taken here is that it permits a thorough analysis of the interactions of a series of different factors that appear to influence character states and thus allows the effects of various evolutionary "decisions" to be examined.

Lovegrove (2009), after denying a potential relationship between reproduction and BMR, concluded that "it remains a great challenge to reconcile ancient physiological adaptation with contemporary adaptation, and in this regard it is imperative that future research focus more on the role of phylogenetic inertia when attempting to explain the significance of physiological diversity." The two most important words here are *ancient* and *inertia*: living organisms must hope that their long-dead ancestors made the right choices because few adjustments of earlier decisions apparently can be made because inertia rules biological existence. Presumably, the ancestors of a species had ancestors, which also had ancestors. So where does "inertia" end? How did this planet obtain such a diversity of life if phylogenetic inertia is so important? This view is essentially antievolutionary in its advocacy of interminable stasis: what ever happened to natural selection? Most closely related species are physiologically similar because they are ecologically and behaviorally similar — that is the principal place where "inertia" comes into play, if indeed this represents inertia — maybe similarity also represents a limit to environmental opportunities. In contrast, ecological and behavioral outliers are usually physiologically distinct, as we have seen. A recent phylogenetic analysis (see chap. 2) of the thermal characteristics of lizards belonging to the genus *Liolaemus* (Labra et al. 2009) "found little evidence for phylogenetic inertia."

The population energetics of marsupials

Marsupial reproduction has profoundly influenced marsupial energetics and ecology. The development of fetal marsupials in utero is greatly restricted compared with that of eutherians (see chap. 13). Furthermore, "the rate of anatomical development after birth during pouch life . . . is also markedly slower for homologous structures than rates seen in [eutherians]" (Lillegraven 1976). In the case of eutherians, the rate of exchange between fetus and mother increases with the mother's rate of metabolism. So, one way to increase the rate of development in mammalian fetuses is to increase the rate of metabolism of pregnant females. Such adjustments are widespread and may be most marked in eutherians and in marsupials with low BMRs (Nicoll & Thompson 1987), and they may partially compensate for low BMRs. But such an adjustment by marsupials is not equivalent to having the high BMRs of eutherians because the adjustment also occurs in eutherians. The time from conception to weaning in marsupials is 50% longer than in eutherians of the same mass (Thompson 1987). Another possibility would be for marsupials to augment the growth of their young by having higher rates of milk production than found in eutherians, but that is also not the case (Thompson 1992). Therefore, the reproductive rates of marsupials are low compared with those of ecologically similar eutherians, except as small marsupials may compensate by increasing litter size. The one circumstance in which marsupials can have the same energy expenditure as eutherians is when both groups have food habits, such as ant and termite eating (see fig. 5.5) and arboreal folivory, that require them to have equally low energy expenditures. As a result of their small range in mass-independent BMR, no clear correlation is found in marsupials among the various parameters of population biology and BMR. However, the restriction of marsupials to low basal rates apparently confines their r_{max} to values that are similar to those of eutherians with equally low basal rates (see fig. 14.8).

Ecological consequences for monotremes and marsupials

The differences in the form of reproduction among monotremes, marsupials, and eutherians have severe consequences for the first two groups. Monotremes, of course, lay eggs. Dependence on egg laying may have restricted the acceptable range of ecological niches for monotremes, although it does not seem to limit the range for birds. The food habits found in monotremes may have protected them from marsupials.

Marsupials pay a high price for their inflexible means of reproduction. The great diversity of marsupials in Australia principally reflects its isolation from the diversity of eutherians. Only two groups of terrestrial eutherians invaded Australia—bats and rodents—which means that many mammalian ecological niches were deprived of eutherians, including those of small terrestrial insectivores, subterranean insect and worm eaters, arboreal folivores and frugivores, carnivores, and large grazers. In their absence, marsupials occupied these niches.

The consequences for marsupials occupying ecological niches that permit eutherians to have high BMRs can be seen in Australia and in South America. Asian traders brought the eutherian wolf, in the form of the dingo, to Australia some 3,500 years ago (Corbett 1995). As a consequence, the two largest carnivorous marsupials, the thylacine (*Thylacinus cynocephalus*), which weighed 20–30 kg, and the Tasmanian devil (*Sarcophilus harrisii*), which weighs 5–12 kg, disappeared, although both species survived in Tasmania, where the dingo never arrived. (Unfortunately, the thylacine in Tasmania was not able to survive the presence of another eutherian predator, *Homo sapiens*—the last one was seen in 1930—and the devil holds on marginally under the threat of a contagious facial tumor disease.) The inability of the thylacine and devil to withstand the presence of the dingo (and people) may have reflected the dingo's superior effectiveness as a predator, but it may also have been due to the dingo's higher energy expenditure and higher reproductive output, which is probably similar to that of the wolf (*Canis lupus*; 128%, whereas the devil has a BMR equal to 87%). A few smaller carnivorous marsupials survive in Australia, particularly those belonging to the genus *Dasyurus*, the largest of which, *D. maculatus*, weighs 2–3 kg and has a basal rate equal to 77% of the value expected from mass. These species survive in the absence of similarly sized eutherian carnivores such as weasels (*Mustela*). Macropods seem to do well in the presence of domestic stock, possibly because of the macropods' ability to survive the absence of surface water, a condition that cannot be tolerated by cattle. How well macropods would survive in the presence of East African ungulates might be a different story.

On island South America, before its attachment via Central America to North America, large mammalian predators belonged to two marsupial families, the Borhyaenidae and the Sparassocynidae. These carnivores disappeared before or with the invasion of South America by eutherian carnivores after its attachment to Central America (Patterson & Pascual 1972; Simpson 1980; Webb 1976, 1978, 1985a, 1991; Marshall et al. 1977, 1979, 1982). Another South American group of carnivores, large flightless, carnivorous "terror" birds belonging to the family Phorusrhacidae, were present in South America (Chiappe & Bertelli 2006; Bertelli et al. 2007). They were able to invade North America and therefore were able to coexist with eutherian carnivores, presumably because they occupied different ecological niches.

The South American endemic marsupials that survived the intercontinental exchange were species adapted to food habits and behaviors associated with intermediate to low mass-independent rates of metabolism. These marsupials included species that are insectivorous, have a mixed diet of fruit and insects, or, as in the case of *Caluromys derbianus*, are arboreal and mix fruit with insects and small vertebrates. No living Neotropical marsupial is committed to a vertebrate diet, the closest being the semiaquatic yapok (*Chironectes minimus*), which feeds variously on crayfish, shrimp, and fish, and *Lutreolina crassicaudata*, which feeds on small vertebrates and insects. These two species have basal rates that are 77% and 93% of the values expected from mass, respectively—values that are typical of marsupials—whereas most Neotropical felids have basal rates that fall between 114% and 131%, although the jaguarundi, a felid somewhat similar to *Lutreolina* in body shape and food habits, has a BMR equal to 80%.

Many low-energy eutherians endemic to South America also survived the exchange, including armadillos, anteaters, and tree sloths. These mammals have low-energy lifestyles that permit them to survive. Most of the high-energy niches in South America are occupied by northern invaders, including felids, canids, mustelids, camelids, and cervids (McNab 1989). The caviomorph rodents endemic to South America are a partial exception, although most have low BMRs (Arends & McNab 2001). Other exceptions include the grazing dasyproctids, which are the ecological equivalent of small bovids or cervids and have BMRs between 106% and 151%, and two aquatic species, the capybara (*Hydrochaeris hydrochaeris*, 123%) and the coypu (*Myocastor coypus*, 208%).

The population energetics of birds

The analysis of eutherian reproduction raises the question of whether a different form of reproduction, the laying of eggs, insulates the population biology of birds from an ecological impact on energetics. If not, why do so many birds have high BMRs? Is it only due to avian flight? Gavrilov (1995, 1998, 1999) argued that one of the factors that led to the high rates of reproduction of temperate passerines was their high basal rates.

Little effort has been expended examining the impact of energetics on avian population biology. In the first attempt to examine this relationship directly, Padley (1985) concluded that the number of eggs per year, incubation time, length of the nestling period, and the postnatal growth constant are independent of BMR in passerines. Unfortunately, mass-specific BMRs were used in this analysis, which, as we have seen, do not separate the influence of body mass from the potentially important influences of ecology and behavior. In a subsequent study, Trevelyan et al. (1990) examined 325 species to determine whether clutch size, incubation period, fledgling period, age at independence, life span, and number of broods per year, among other variables, are correlated with mass-independent basal rate. They found no evidence of such correlations, although some question exists as to the quality of the rates of metabolism used, since some of the measurements were made "without defining the thermoneutral zone." Furthermore, the analysis was based on the Sibley-Ahlquist cladogram.

In an extensive study of the postnatal exponential growth constant k in temperate and tropical birds, Ricklefs (1968, 1976) concluded that growth constants were lower in tropical species. (Ricklefs [1968, 1976], followed by Drent and Klaassen [1989], referred to the growth *constants* as growth *rates*. If rates, they would have had units of mass/time, but they actually had units of 1/time, i.e., a constant [see eq. 14.3]). This finding raises the possibility that the lower constants in tropical species might reflect the low mass-independent BMRs in tropical species (see chap. 3). To test this possibility, 130 values for k were extracted from Ricklefs; they correlated principally with the growth asymptote of body mass (fig. 14.10; $P < 0.0001$),

$$k = 1.22 \ m^{-0.369} \tag{14.7}$$

($r^2 = 0.715$), as well as with the asymptote ($P < 0.0001$) and climate ($P < 0.0005$),

$$k = 1.47 \ (C) \ m^{-0.393} \tag{14.8}$$

($r^2 = 0.741$), where the climate coefficient C equals 0.75 in tropical species and 1.00 in temperate species (four species were polar, but they did not differ from temperate species, so they were combined with the temperate

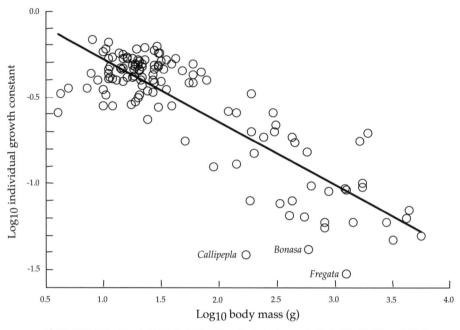

Figure 14.10. Log$_{10}$ growth constant as a function of log$_{10}$ body mass in birds. (Modified from Ricklefs 1976 and Drent & Klaassen 1989.)

species). That is, the 40 tropical species had individual growth constants that averaged 75% of those of the 90 temperate species, a finding that agrees with Ricklefs's (1976) conclusion that tropical species have lower growth rates. The species with very low growth constants included *Callipepla californica* and *Bonasa umbellus*, neither of which have had their basal rates measured, but both of which belong to families that usually have low BMRs.

Did the lower growth constants among tropical species reflect lower BMRs? Estimates of BMR were available for only 43 of the 130 species. Mass-independent growth constants, derived from the general mass growth curve (eq. [14.7]), were not significantly correlated ($P = 0.86$) with mass-independent BMRs, even when climate was added as factor ($P = 0.56$). One of the potential difficulties in this analysis is that there was little variation in mass-independent BMR, 25 of the species having values between 80% and 110%. An appreciable residual variation in k was unaccounted for, possibly in relation to other factors. If so, those factors did not include food habits ($P = 0.80$).

Maybe this analysis tried to make too broad a comparison, because some taxonomically limited comparisons demonstrated a correlation of the variation in mass-independent growth constants and mass-independent BMRs (fig. 14.11). For example, estimates of the mass-independent growth constants of juvenile gulls and terns, procellariiforms, and shorebirds each correlated with mass-independent rates of metabolism (Drent & Klaassen 1989). The only hesitancy is whether the energy expenditures measured were really *basal*, because the birds were juveniles and were undoubtedly not postabsorptive, although they were apparently measured in thermoneutrality. The other question is what standards should be used to judge the level of the reported BMRs and growth constants. Drent and Klaassen apparently used internal standards for each of the three groups of birds. However, even if one uses the general bird curve (eq. [1.5]) for BMR and the bird growth curve (eq. [14.7]) for k, there remains a slightly chaotic dependence of k (%) on BMR (%). Two gulls (*Larus ridibundus* and *L. atricilla*) have unusually low ks, as was noted by Drent and Klaassen (1989) and by Dawson et al. (1976), as do two albatrosses (*Diomedea*) (see fig. 14.11). Tentatively, the individual growth constant k appears to depend on BMR in some birds.

Another possibility is that the annual egg production of birds correlates with energy expenditure. Annual egg production is a product of mean clutch size and annual clutch frequency. Among closely related species, clutch size generally increases with latitude (Moreau 1944; Lack 1947, 1948; Skutch 1949; Cody 1966; Yom-Tov 1987, 1994; Martin 1996; Martin et al. 2000;

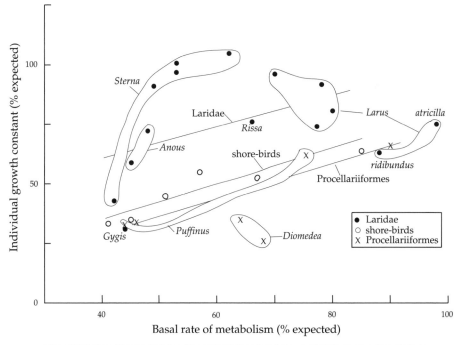

Figure 14.11. Mass-independent growth constants for related groups of birds as a function of mass-independent basal rate of metabolism. (Data from Ricklefs 1976 and Drent & Klaassen 1989.)

Patten 2007; McNamara et al. 2008): Neotropical birds usually have clutches of 2 or 3 eggs, whereas north-temperate species usually have 4 or 5 eggs (Martin 1996), and Arctic species have 5 to 7 eggs (Hussell 1972; Ricklefs 1980). Various explanations have been given for this correlation, including compensation for mortality, climate stability, nest predation, food delivery rate, and phylogeny, among others—all of which deal with the rationale for an increase or decrease, not with the mechanism. Clutch size in South African birds correlates with rainfall seasonality and stochasticity (Lloyd 1999; Lepage & Lloyd 2004), and clutch size is reduced in regions with a stable climate, such as the tropics, oceanic islands, and coastal areas of continents (Cody 1966). That is, clutch size in birds is environmentally sensitive.

The geographic dependency of clutch size has not been examined with respect to energetics. Clutch size was suggested by Moreau (1944) to reflect mortality, or is this correlation reversible? The large clutch sizes of temperate and polar species may compensate for high mortality associated with generally higher mobility and migratory behavior, whereas the small clutch sizes of tropical species may be associated with their sedentary habits and reduced mortality. The small clutch size of birds endemic to New Zealand and Australia (Franklin & Wilson 2003) may also reflect a sedentary existence, which surely is the case for New Zealand endemics. We have already seen that both flighted and flightless ducks endemic to New Zealand are sedentary and are characterized by low mass-independent BMRs (McNab 2003a). The mean clutch size of species resident in Israel is lower (3.93) than that of migrants (4.45), a pattern that Yom-Tov (1994) suggested may be associated with the loss of migrants traveling across the Sahara. Furthermore, birds of the Cape Province of South Africa have a lower mean clutch size than birds in Israel, which correlates with a smaller proportion of migratory species (Yom-Tov 1994).

If clutch size is larger in migratory species, then the distribution of clutch sizes also correlates with that of mass-independent BMR because both are associated with the migratory/sedentary dichotomy. Whether this correlation implies that a direct connection exists between clutch size and mass-independent BMR is unclear. The correlation of clutch size with latitude is unlikely to be due to latitude per se, or to any of its physical correlates, such as photoperiod or ambient temperature. A more direct connection between clutch size and basal rate is found in rails, in which flightless species have both lower mass-independent BMRs and smaller clutches than volant species (see fig. 9.7; McNab & Ellis 2006). Whether this correlation is determinative is unclear.

Of course, another way to modify annual egg production is to vary the number of clutches produced in a season. There is great variation in brood numbers and renesting rates in most environments, and these parameters may not differ between geographic and climatic areas (Martin 1996). However, a recent analysis (Tieleman et al. 2006) of FEE and clutch size in temperate and tropical populations of the house wren (*Troglodytes aedon*) indicated that an Ohio population had a mean FEE that was 2.9 times the basal rate expected from equation (1.5) and a clutch size of 6.0–7.5, whereas a Panamanian population had a FEE that was 2.1 times the expected basal rate and a clutch size of 3.5–3.6, one-half that of the temperate, migratory population.

Tentative conclusions

Some limited conclusions about the relationship between energetics and population biology in endotherms can be tentatively made. Eutherians have a form of reproduction that facilitates the influence of various ecological factors on energy expenditure. This coupling of the parameters of reproduction to energy expenditure permits eutherians to be effective exploiters of opportunities present in the environment. Marsupial reproduction, however, does not permit the effective exploitation of high-quality resources because of the narrow variation of mass-independent energy expenditure dictated by the characteristics of marsupial reproduction. Some habits or environmental conditions appear to depress eutherian energy expenditures to the level found in marsupials with the same habits. The only surviving monotremes occupy ecological niches that require some of the lowest energy expenditures. The relationship of population biology to energetics in birds is much less clear. Some temperate and polar birds have high mobility, high reproductive rates, and high mass-independent BMRs, whereas many tropical species have a sedentary lifestyle and low mass-independent BMRs. However, the extent to which clutch size, mobility, mortality, and mass-independent BMR are functionally connected in birds is unknown, a situation that contrasts with the linkage seen in eutherians.

CHAPTER FIFTEEN. *The Evolution of Endothermy*

A fundamental historical breakthrough occurred with the evolution of endothermy in the classes Aves and Mammalia. Endothermy permitted birds and mammals to occupy many more ecological niches than are available to ectothermic vertebrates, especially in terrestrial environments. Ectothermic vertebrates are usually excluded from cold-temperate and polar terrestrial climates, except for a few small amphibians and reptiles. How is it that evolution could facilitate the radical transformation of ectothermy into endothermy? This question can be answered at the molecular and organ system levels, or at the level of the intact physiological function of an individual (Ruben 1995; Kemp 2006). At whatever level one considers this question, a major change appears to have involved a great increase in energy expenditure. Given the independent evolution of these two classes, what is most striking is how similar their endothermies are. That similarity clearly implies that the physical principles controlling heat exchange and the maintenance of temperature differentials (see chap. 1) determine the general parameters of both avian and mammalian endothermy (Gates 1962, 1980), although some biologically important adjustments have occurred. We have examined some of the factors responsible for the modest difference between avian and mammalian endothermy in chapters 1, 3, 10, 11, 13, and 14.

Mammals have *basal* rates of metabolism that are about 34 times the standard rates of reptiles with a body temperature of 20°C and 12 times those of reptiles with a body temperature of 30°C (McNab 2002a). Birds have basal rates that are about 68 times the standard rate of reptiles with a body temperature of 20°C and 25 times those of lizards at 30°C. These ratios are based on the coefficients for the metabolism-mass curves, which is possible because they have similar powers of body mass. A better comparison of reptiles and mammals might be of field energy expenditures when reptiles have attained their preferred body temperatures. Then, mammals have field energy expenditures that are about 3 times their own basal rates, about 17 times the field energy expenditures of reptiles at a mass of 10 g, and about

8 times the field energy expenditures of reptiles at 1 kg (Nagy et al. 1999). As a concrete example, the field energy expenditure of a 43.3 kg African spring-bok (*Antidorcas marsupialis*) was about 10 times that of a 45.2 kg Komodo monitor (Nagy et al. 1999). The shift from the ectothermy of reptiles to the endothermy of birds and mammals therefore involved an appreciable in-crease in energy expenditure, although this came about in mammals, para-doxically, in association with a *decrease* in energy expenditure.

The evolution of endothermy in mammals

A variety of factors have been suggested to be related to the shift from the ec-tothermy of reptiles to the endothermy of mammals, including a change in limb-supported posture (Heath 1968), mammalian reproduction (Hopson 1973), a reduction in body size (Hopson 1973; McNab 1978a), formation of a secondary palate (McNab 1978a), bone morphology (de Ricqlès 1974), predator/prey ratios (Bakker 1975, 1986), the occupation of a nocturnal niche (Crompton et al. 1978), a metabolic response to activity (Taigen 1983), an increase in aerobic capacity (Hayes & Garland 1995; Seebacher et al. 2006; Hayes 2010), the evolution of maxilloturbinals (Hillenius 1994; Ruben 1995; Hillenius & Ruben 2004), and the integration of most, or all, of these factors (Kemp 2005, 2006). All of these factors either contributed to or resulted from the evolution of endothermy in the phylogeny of mammals.

A factor that must be taken into consideration is the appreciable decrease in body size that occurred in the evolution of mammals from therapsid rep-tiles. This decrease in mass may have been by as much as a thousandfold— that is, possibly from 5–10 kg to 5–10 g (McNab 1978a)! Any appreciable decrease in mass is important because mass is always the most important factor determining energy expenditure as long as mass range is appreciable. A decrease in mass by a factor of 1,000 implies a very large *decrease* in en-ergy expenditure, by a factor of about $(1,000^{0.75})^{-1}$—that is, to a rate that is only 0.56% of the original rate—irrespective of any shift from ectothermy to endothermy.

What is most important in this discussion is the distinction between *total* and *mass-specific* rates of energy expenditure. This distinction is important because it can be startling to hear that the evolution of mammalian endo-thermy was principally associated with a *decrease* in energy expenditure, but that decrease was in total rate of metabolism, which is the *effective*, ecologi-cal, evolutionary rate. The principal way in which the evolution of endo-thermy could involve an *increase* in total rate would be if, say, a 20 g reptile gave rise to a 20 g mammal, which of course did not occur.

In a more recent evaluation of the decrease in mass associated with the evolution of the first mammals, Kemp (2005) noted that many of the lines leading to mammals had a diversity in body mass, but in the final analysis, "the earliest mammals, which were very small, shrew-sized mammals," might have weighed 5–15 g. To continue this argument, mass is proportional to a volume, which is estimated by the cube of a linear measure multiplied by a density. Thus, if skull length decreased from 200 mm to 20 mm during the evolution of mammals, the estimated mass decrease would equal approximately $(20/200)^3 = 0.001$; that is, to one-thousandth of the original mass. This decrease might be exaggerated, but even if the reduction was from 3 kg to 7.5 g, such a 400:1 reduction would have led to a decrease in rate of metabolism to approximately $(1/400)^{0.82} = 0.0074$ (0.82 being the power of lizard rates at $T_b = 30°C$ [Bennett & Dawson 1976]). Then total rate of metabolism would have been only 0.74% of the original therapsid value, again assuming no shift to an endothermic curve. If the endothermic curve was 10 times that of a 30°C lizard curve, then the rate of metabolism of the first (endothermic) mammal would have been roughly $0.74(10) = 7.4\%$ of the original therapsid value (fig. 15.1). Clearly, the decrease in mass was the single most important factor in shifting the rate of metabolism of a large ectotherm to that of a small endotherm.

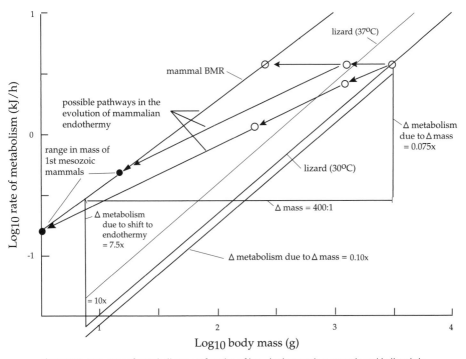

Figure 15.1. Log_{10} rate of metabolism as a function of log_{10} body mass in mammals and in lizards in relation to the evolution of endothermy in mammals. (Modified from McNab 1978a.)

This analysis neglects the question of whether the first mammals were persistently endothermic, which, as we saw in chapter 4, requires even higher basal rates than found on the standard mammalian curve at masses <45 g. For example, the first shrew-sized mammals may have been physiologically more similar to crocidurine shrews, or to dasyurine marsupials, than to soricine shrews in that they entered torpor. The transition from ectothermy to endothermy was probably gradual (see fig. 15.1), so that some of this shift may have started with the first reduction in size and continued in fits and starts, depending on the various ecological and behavioral characteristics of transitional species, as the decrease in mass continued to the earliest mammals.

Kemp (2005, 2006) suggested that this analysis represents a "thermoregulation-first" explanation, which is incorrect; it simply represents the view that energetics was an important *component* in the evolution of endothermy. The adjustment of a large number of ecological and behavioral factors would be included in the evolution of endothermy as mass decreased, and these factors may have actually been the impetus to convert ectothermy to endothermy. The attempt to make this conversion simply by overfeeding a lizard (Bennett et al. 2000) is bizarre: the actual evolution of endothermy was undoubtedly physiologically and ecologically infinitely more complex.

A *gradual* evolution of endothermy in the phylogeny of mammals is consistent with the observation that the evolution of a secondary palate first occurred in the dicynodonts, which suggests that the shift from an ectothermic to an endothermic metabolism curve started long before the first mammals appeared (McNab 1978a). A secondary palate facilitates an increased rate of ventilation, associated with the shift in metabolism, while buccal activities, such as eating, were increasing. Consequently, the dicynodonts probably fell between the ectothermic and endothermic metabolism curves, with the contribution of heat production to temperature maintenance increasing with the reduction in body mass (see fig. 15.1). Furthermore, the development in some therapsids of maxilloturbinals undoubtedly facilitated higher rates of gas exchange and a reduction in respiratory water loss (Hillenius 1994; Ruben 1995). Furthermore, Nespolo and colleagues (2011) argued that if maximal energy expenditure is connected to the standard rate of metabolism, an increase in aerobic capacity might contribute to the evolution of endothermy. In this view, the shift from ectothermy to endothermy resulted from selection for an increase in the maximal rate of metabolism. The shift from ectothermy to endothermy continued with a further shift in the galesaurids, and was possibly completed in the first mammals. All of these changes were occurring as body mass decreased. Somewhere in this process the development of a fur coat occurred, presumably gradually as the rate of metabolism shifted from a reptilian to a mammalian level as mass deceased.

The evolution of endothermy in birds

Almost nothing is known of the evolution of endothermy in the class Aves, principally because of the question as to the origin of birds. The majority of vertebrate paleontologists apparently believe that birds were derived from coerulosaur dinosaurs. However, some ornithologists, notably Alan Feduccia (1996), have questioned this conclusion. He and his colleagues have suggested that all feathered "dinosaurs" really were birds (Feduccia et al. 2005). Even if birds were derived from dinosaurs, the detailed sequence by which bird evolution occurred is unknown. That leaves the question of the origin of avian endothermy up in the air because one cannot analyze anatomical changes that occurred in the evolutionary sequence. Although there has been much speculation on the thermal biology of dinosaurs (Bakker 1980; Ostrom 1980; Weaver 1983; Paul 1988; Falow et al. 1995; Ruben et al. 1996; Seebacher et al. 1999; Ruben et al. 1997; Sander & Clauss 2008; McNab 2009c), it is unlikely to have been uniform because of the great diversity in their mass. The presence of endothermy in the largest species is unlikely (McNab 2009c). One fascinating aspect of the dinosaurian hypothesis for the origin of birds is that birds would be a small derivative of dinosaurs, and therefore may have evolved endothermy in association with a large reduction in body mass and energy expenditure, as has been suggested in the phylogeny of mammals.

CHAPTER SIXTEEN. *The Restrictions and Liberations of History*

Evolutionary adjustments in energetics can restrict or expand future evolutionary and ecological opportunities. Some character states may represent advancements reflecting the immediacy of environmental opportunities in the short term, but appear over a longer period to restrict an organism's capacity to face future challenges and exploit new opportunities. These character states in mammals include a commitment to a monotreme or marsupial form of reproduction, a fossorial (subterranean) existence, an arboreal lifestyle, or an armored condition, whereas in birds the evolution of a flightless condition constitutes such a restriction. In contrast, a series of evolutionary innovations permitted some vertebrates to expand their temporal and geographic ranges and dominate many niches. These evolutionary accomplishments include the independent attainment of endothermy by mammals and birds, the mammalian invasion of the marine environment, the eutherian form of reproduction, and the attainment of flight. These restrictions and liberations are another way in which history (phylogeny) has had an impact on contemporary character states and their consequences.

Restrictions

A monotreme or marsupial form of reproduction

As we have seen (in chaps. 13 and 14), marsupials have a form of reproduction that does not permit them to exploit fully the high-energy resources in the environment, and the same situation undoubtedly exists in monotremes. Marsupials are at a disadvantage in competition with eutherians, at least in the exploitation of high-quality resources, although they can coexist with eutherians if both groups exploit marginal resources that require all consumers to reduce their energy expenditures. Historically, marsupials have successfully exploited high-quality resources only in the absence of ecologically equivalent eutherians, and even then, marsupials do not have high energy expenditures. When eutherians that exploit high-quality resources came into contact with marsupials occupying the same niches, the marsupials invariably lost the competition—a pattern that occurred in Australia and South America—because the high rates of metabolism in eutherians are translated into high rates of reproduction (see chap. 14). Marsupials, in turn, may have an advantage over monotremes except in habitats that require the lowest energy expenditures, which may explain why monotremes selectively survived in Tasmania, Australia, and New Guinea. The fossil record of monotremes is very poor, but all known species occupied low-energy habitats, including Cretaceous species in both Australia and South America (Archer et al. 1985; Pascual et al. 1992; Flannery et al. 1995; Pascual et al. 2002).

A fossorial life

Fossorial mammals make many adjustments to a permanent underground life, including reduction of energy expenditures (see chap. 8). However, the potentially most radical adjustment made by some fossorial mammals is a reduction in vision in response to a reduction of ambient light (Burda 2006). This adjustment has reached its extreme in a blind condition, with degenerate eyes covered by skin, in *Spalax* and some moles. Such an extreme adjustment, possibly more than any other, represents a commitment to a lifestyle that is difficult, and probably impossible, to reverse.

A mammalian arboreal life

Many mammals are committed to arboreal lifestyles, especially in tropical forests (see chap. 8). Most arboreal mammals fall into one of two categories: highly active or sedentary. Some sedentary species, including sloths and phalangers, feed on leaves; others feed on insects—namely, ant and termite eaters and some smaller primates (see chap. 8). These species have BMRs that range between 35% and 90% of the values expected from mass. Intermediate levels of activity exist; an example is the colobus monkey (*Colobus* sp.), a semi-inactive species that feeds on leaves and has an intermediate BMR (108%). The radical reduction in BMR and the associated reduction in temperature regulation in sedentary arboreal species prevents them from entering temperate environments and often excludes them even from cool tropical cloud forests (see chap. 12).

An armored life

Several groups of mammals evolved cutaneous armor. A question that armor raises is whether its evolution is related to energy expenditure. Armored monotremes (*Tachyglossus* and *Zaglossus*), which feed on ants and termites or on worms, have basal rates that are between 32% and 45% of the values expected from mass (McNab 1984). Nine species of armadillos (Dasypodidae), all of which are armored, weigh between 1 and 45 kg and have basal rates between 37% and 66% (McNab 1980c). The two armadillos in this group with the lowest BMRs, *Tolypeutes matacus* at 1 kg (37%) and *Priodontes maximus* at 45 kg (38%), are committed to ant and termite eating. Another group of armored anteaters, the pangolins (Pholidota, Manidae), have BMRs between 31% and 86% (Hildwein 1972, 1974; Heath & Hammel 1986; McNab 1984).

Two families of porcupines exist: the Hystricidae, which is limited to Eurasia and Africa, and the Erethizontidae, found in the Americas. The few porcupines measured include the Cape porcupine (*Hystrix africaeaustralis*), which belongs to the Hystricidae and has a BMR equal to 73% (Haim et al. 1990), and two species that belong to the Erethizontidae: the Neotropical *Coendou prehensilis*, which is arboreal and has a BMR equal to 77% (McNab 1978c), and the temperate North American *Erethizon dorsatum*, which is semiarboreal and has a BMR equal to 97% (Arends & McNab 2001). In the case of *E. dorsatum*, climate is a potential factor influencing its basal rate.

The Eurasian and African hedgehogs (Erinaceidae) and the Madagascan tenrecs (Tenrecidae) are especially interesting from the viewpoint of energetics because both families have spiny and non-spiny species. The Erinaceidae is divided into two subfamilies, the Erinaceinae, the typical hedgehogs with well-developed spiny coats, and the Galericinae, which do not have spiny coats. Five species of typical hedgehogs have been measured, including species belonging to the genera *Hemiechinus*, *Paraechinus*, *Atelerix*, and *Erinaceus*, which collectively have BMRs that vary between 39% and 82%. The highest values (79% and 82%) are found in the two European species of *Erinaceus* (Shkolnik and Schmidt-Nielsen 1976; McNab 1980c; Król 1994), which may reflect their occurrence in seasonally cold environments, although both species hibernate in winter. In contrast, two species of Galericinae, the lesser gymnure (*Hylomys suillus*) and the moon rat or gymnure (*Echinosorex gymnurus*), both from Southeast Asia, lack spiny coats and have BMRs of 100% and 126%, respectively (Whittow et al. 1977; Genoud & Ruedi 1996). Thus, within this family, lower BMRs correlate with the occurrence of spiny coats, whereas higher BMRs correlate with furry coats, even though the gymnures live in tropical climates.

Tenrecs (Tenrecidae) that are members of the Tenrecinae have spiny coats and have basal rates that vary between 38% and 74% (Hildwein 1970; McNab 1980c; Stephenson & Racey 1994), whereas members of the Oryzorictinae have fur coats and basal rates between 78% and 144% (Stephenson & Racey 1993b; Stephenson 1994a,b). *Geogale aurita*, a tenrec that belongs in its own subfamily, lives in dry environments, has a fur coat, and has a basal rate of 57% (Stephenson & Racey 1993a). Almost all tenrecs feed on soil insects and earthworms. With the exception of *Geogale*, armored species in this family have lower BMRs than furred species; the low BMR in *Geogale* may reflect its dry environment.

The association of an armored integument with a reduced basal rate re-emphasizes the caution previously given that correlation does not necessarily imply cause and effect. A direct connection between these characters is unlikely, but that does not mean that they are unrelated. The most likely explanation for the depressed level of energy expenditure in armored species is that species that search for food in the soil (hedgehogs, tenrecs, armadillos, pangolins) and move slowly have behaviors that make them vulnerable to predators. In response, these species evolved passive integumental devices to reduce their vulnerability. These species often take a defensive posture by crouching on, or digging into, the ground or by contracting into a ball. Relatives that do not have spiny coats or armored plates are likely to be more agile, evading predation by scurrying away.

Lovegrove (2001) performed a general analysis of the occurrence of armor in mammals. He associated armor with feeding on limited resources, a limited capacity for rapid movement, and low basal rates of metabolism, a view similar to what is argued here. He also related it to a concept of a limited range of opportunities at an intermediate mass of about 358 g, above and below which a greater range of mass-independent basal rates occurs. The greater range of mass-independent basal rates at smaller masses is associated with the use (or not) of torpor (see chap. 4), and the greater range at larger masses reflects the differential use of marginal or high-quality food resources or life in restrictive or expansive environmental conditions.

A flightless condition in birds

One of the most remarkable evolutionary adjustments to environmental conditions is the repeated evolution of a flightless state in birds on oceanic islands (see chap. 10). Before the arrival of people, some thirty to thirty-five species or subspecies of terrestrial and freshwater birds in New Zealand, or 25% to 35% of the endemic avifauna, were flightless. Today, greater New Zealand has only three flightless kiwis, two flightless ducks, two flightless rails, and a flightless parrot. At least twenty species, or approximately 24% of Hawaiian endemic terrestrial and freshwater avifauna, were flightless (McNab 1994a). Now all flightless birds of Hawaii are extinct. Other living flightless birds include the kagu (*Rhynochetos*) of New Caledonia, a flightless cormorant (*Phalacrocorax*) in the Galápagos, a flightless rail in New Guinea and perhaps fifteen or so on other islands of the South Pacific, a flightless rail on Aldabra, which is the last flightless bird in the Indian Ocean, and another on Inaccessible Island in the South Atlantic.

A flightless condition in birds makes them extremely vulnerable to predators, and its evolution normally occurs only in the absence of eutherian predators. Part of the difficulty faced by flightless birds is that they have low rates of reproduction (McNab & Ellis 2006). Six flightless rails have a mean BMR that is 27.5% lower than the mean BMR of eight flighted rails. Clutch size in rails decreases in proportion to the decrease in mass-independent BMR (see fig. 9.7). Consequently, flightless rails are susceptible to eutherian predators because they cannot evade predation and they cannot compensate for the increase in mortality through an increased reproductive output.

As a result of this extreme adjustment to an island life, flightless birds in general are doomed as a result of the activities of *Homo sapiens*, unless they are the size of rheas, cassowaries, emus, or ostriches, which can outrun people and other predators. Steadman (2006) stated that "my interpretation of the fossil record and historic specimens roughly suggests that from 500 to 1600 species of flightless rails inhabited Pacific islands at human contact. Rails alone account for most of the roughly 1000 to 2000 species of birds that I believe would exist today had people never colonized Oceania . . . . If not for human impact, more species of rails would be alive today than of any other family of birds. What a shame that 98% or 99% of these rails are gone." Flightless bird species are likely to survive only as long as their islands remain free of eutherian predators and above water (see chap. 17).

Is a commitment to a restrictive lifestyle reversible?

The evolution of a commitment to a restrictive lifestyle usually involves substantial modifications of morphological characters in addition to a reduction in energy expenditure. Once such commitments have been made, is it possible, under appropriate conditions, to reverse those modifications to the point that the descendants of the modified species could return to a less restrictive lifestyle? In other words, could fossorial lineages live above ground, arboreal species become terrestrial, or flightless birds become flighted? Or do such extreme modifications represent terminal commitments? If some of these phenotypic losses occur because some genes are "turned off," can they be "turned on" again? Or is the genetic basis for these adjustments so complicated that there is little likelihood that the ancestral condition can be reconstructed?

Among vertebrates, a flighted condition has evolved from a flightless condition at least three times, in pterosaurs, birds, and bats. But has flight ever evolved secondarily from a flightless condition that itself evolved from a flighted condition? Such re-evolutions appear to be unlikely, but an interesting possible exception is the flighted condition in tinamous, the only living birds belonging to the ratite assemblage that are capable of flight. The flight apparatus in tinamous is morphologically different from that found in most flighted birds in that they have very large flight muscles (29%–40% of total mass) with no muscle myoglobin, they tire rapidly in flight, they have small lungs, and they have proportionally the smallest heart of all birds, but they have a keeled sternum (Sick 1993).

DNA base sequences (Hackett et al. 2008; Harshman et al. 2008) and mitochondrial genome sequences (Phillips et al. 2010) indicate that tinamous evolved from the middle of the ratites; that is, after the lines that led to ostriches and rheas, but before cassowaries, emus, and kiwis (fig. 16.1). Does that mean that the flightless condition in ostriches and rheas evolved after the line leading to tinamous was separated, with the tinamous retaining an early flighted condition? If so, a flightless condition in this order would have had to evolve three or more times. Or does it mean that the flighted condition in tinamous re-evolved from an early flightless condition in this order? Harshman et al. (2008) and Phillips et al. (2010) preferred the view that tinamous retained their flight from earlier flighted ancestors and that the evolution of a flightless condition in ratites occurred independently, a pattern that is most likely. Otherwise, the distinctive morphological characteristics of the flight apparatus of tinamous represent a secondary evolution of flight. If the flighted condition of tinamous represents a retention from the early ancestors of birds, what were those ancestors?

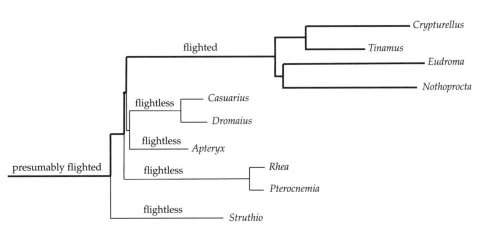

Figure 16.1. Suggested evolutionary history of ratites in relation to the occurrence of a flighted or flightless condition. The most likely pathway of the occurrence of a flighted condition leading to that of tinamous is indicated by the darkened pathway. (Modified from Harshman et al. 2008.)

Liberations

As noted, some evolutionary changes have expanded ecological opportunities for birds and mammals. These changes include the evolution of endothermy, mammalian invasions of the marine environment, eutherian reproduction, and avian flight.

The evolution of endothermy

The evolution of endothermy permitted birds and mammals to occupy essentially all terrestrial environments. The low ambient temperatures at high latitudes and high altitudes are no serious impediment to activity as long as species have a sufficient level of energy expenditure, enough insulation to tolerate large—often huge—temperature differentials, and sufficient food to pay the cost of endothermy, or at least have the ability to evade seasonally harsh conditions.

Some large terrestrial Arctic and alpine species, such as mountain goats (*Oreamnos americanus*), caribou (*Rangifer tarandus*), musk oxen (*Ovibos moschatus*), and wolves (*Canis lupus*), can maintain temperature differentials that may exceed 80°C! Various shrews and arvicoline rodents remain active throughout winter in the Arctic, although they often reduce the temperature differential by using sheltered runways, thereby reducing energy expenditure. However, the collared lemming (*Dicrostonyx groenlandicus*), at 60 g, spends much of its time on the snow surface, as suggested by its white winter coat, cryptic against the snow. Most birds that are summer residents of the Arctic, especially those that require open water, migrate to warmer climates for winter, thereby avoiding the coldest temperatures, but a few, such as ptarmigans (*Lagopus* spp.), gyrfalcons (*Falco rusticolus*), the snowy owl (*Nyctea scandiaca*), the hawk owl (*Surnia ulula*), the northern raven (*Corvus corax*), and a few finches (e.g., hoary redpoll [*Carduelis hornemanni*], pine grosbeak [*Pinicola enucleator*], and bullfinch [*Pyrrhula pyrrhula*]) usually remain in the far north throughout the year. In winter, the predatory species feed principally on lemmings or other arvicolid rodents, whereas the finches depend on seeds. Terrestrial reptiles, in contrast, are excluded from latitudes much above the U.S.–Canadian border, except near the Great Lakes and along the Atlantic coast. The only reptile that has an appreciable distribution in Canada is the common garter snake (*Thamnophis sirtalis*), which is found at 60° N—close to the northern edge of Saskatchewan—but spends much of the year in torpor.

The invasion of aquatic environments

Endothermy contributed to the ability of mammals to invade aquatic environments. Such invasions have occurred many times, and the invaders have included cetaceans, seals, dugongs, mustelids, and (nearly unbelievably) even sloths (de Muizon & McDonald 1995; de Muizon et al. 2004; Pujos & Salas 2004)! A marine environment is nutritionally much richer than a terrestrial environment, which is why marine mammals attained a much larger maximal size than any terrestrial animal, including the largest dinosaurs (McNab 2009c). Mammals became the largest herbivores (e.g., blue whale [*Balaenoptera musculus*], 150 t; bowhead [*Balaena mysticetus*], 60 t) and carnivores (e.g., sperm whale [*Physeter catodon*], 40 t; orca [*Orcinus orca*], 9 t) in the marine environment. Some large ectothermic vertebrates are also present there, including the basking shark (*Cetorhinus maximus*, 4 t) and whale shark (*Rhincodon typus*, 14 t), both of which, like the large baleen whales, are filter feeders. Some of the smaller sharks and predatory teleosts have evolved endothermy. The birds that are most committed to a marine existence are penguins, which have given up flight and spend much of their life immersed in South Polar waters.

Eutherian reproduction

As we saw in chapters 13 and 14, the eutherian form of reproduction permitted eutherians to dominate high-energy food sources, especially grass and vertebrates, as long as they have an opportunity to get to a location. Eutherians dominated carnivory in the Northern Hemisphere but were excluded from land masses in the Southern Hemisphere until the establishment of the Central American land bridge and the approach of Africa to southern Eurasia. South America, for example, was isolated from North America until the mid-Miocene (ca. 20 mya), when there was a limited faunal exchange, which became a major exchange by 3.5 mya. At that time, eutherian carnivores entered South America, and a faunal restructuring began with the extinction of its marsupial carnivores. Australia, however, was never connected to Holarctica, which is why grazing and carnivorous niches remained occupied by marsupials.

Flight

One of the most remarkable evolutionary accomplishments by vertebrates was the evolution of flight. Flight led to the geographic dispersion of birds to every part of this planet. As a result, avian diversity dominates the North

Pole on a seasonal basis, and birds are the dominant vertebrates on oceanic islands. Indeed, oceanic islands essentially became bird paradises. Mammals were generally absent, except for a few bats. As a result, large oceanic islands and archipelagos, such as New Zealand, Hawaii, the Solomon Islands, New Caledonia, Fiji, Samoa, and the Mascarene Islands, evolved complex and distinctive avian communities. This situation applied even to large islands such as New Guinea and those in the Bismarck Archipelago, as well as to Australia and Tasmania, whose terrestrial eutherian fauna is limited to rodents and bats. Various ecological niches that were occupied in Holarctica by eutherians were occupied on these islands by birds, including grazing and browsing by elephant birds in Madagascar and moas, the kakapo (*Strigops habroptilus*), and the takahe (*Porphyrio mantelli*) in New Zealand, and feeding on soil invertebrates by kiwis in New Zealand.

Bats also evolved flight. Their flight is more restrictive than that of birds, possibly because of the earlier presence of birds. Bats are essentially limited to nocturnal habits and to a rather restricted set of food resources, principally fruit, nectar, and flying insects, although three vampire bats occur in the Neotropics and some (few) "insectivorous" bats acquired a taste for small mammals and birds. Insectivory and frugivory have generally restricted the distributions of bats. The only food resource widely used by bats in temperate regions is flying insects, which precludes year-round activity; the bats' usual response is hibernation in caves. The New Zealand bats that belong to the Mystacinidae, which feed on insects, fruit, and nectar, are among the few exceptions (Daniel 1979; Arkins et al. 1999). In the Old World tropics and on islands in the South Pacific, megachiropterans (flying foxes) have occupied the nocturnal frugivorous niche.

Long-term survival in low-energy niches

Some living endotherms retain characteristics that reflect an early stage of evolutionary development. Among mammals, three types of monotremes survive in Australia and New Guinea, a variety of marsupials exist in the New World, and a greater ecological diversity of marsupials occurs in Australia and New Guinea. But even within the Eutheria, a variety of "nonprogressive" ecotypes persist, including pangolins in Asia and Africa and tamanduas, armadillos, and sloths in the New World. Some living eutherians represent the last remnants of formerly more widely distributed groups, such as lemurs in Madagascar.

What is most interesting about this list is that all these species have low mass-independent BMRs. None occupy high-energy niches—carnivory and grazing—except in the absence of ecologically similar eutherians. The protection afforded these relicts is that they exploit resources unacceptable to high-energy eutherians. Thus, all ant and termite eaters that weigh more than 100 g have low BMRs, irrespective of whether they are monotremes, marsupials, or eutherians (McNab 1984). Some birds also are evolutionary relicts, especially the ratites of Africa, South America, New Zealand, Australia, and New Guinea; all measured species have low basal rates.

Historical protection appears to have been given to species that exploit resources that require a low level of energy expenditure because these niches diminish any advantage that would accrue to potential competitors from an increased rate of reproduction derived from a high rate of energy expenditure (McNab 2006b). The impact of history, therefore, is not uniform, but selective. However, this type of long-term persistence has a weakness: in the presence of increased mortality, which is often associated with the presence of humans, species with low rates of reproduction are the most vulnerable to extinction because they cannot compensate with an increase in fecundity.

CHAPTER SEVENTEEN. *Global Issues:*
The Limitation to a Long-Term Future

One of the privileges of being a biologist is the opportunity to study the great range of organisms that are found on Planet Earth. A great sadness is realizing how many species have been lost as a result of human activity. We have lost thousands of birds endemic to oceanic islands, untold numbers of South American endemic mammals, north-temperate giants, and Australian endemics, and the African megafauna is threatened today. Various factors have contributed to these losses and threaten species now and in the future. The principal factor is the activities of humans, including our destruction of natural environments, our manufacture and use of chemicals, which often turn out to be toxic to other organisms, and our contributions to global warming, as well as the direct killing of animals. All these phenomena ultimately result from the exponential increase in the world's human population. A recognition of this human impact was codified in an informal geological period, the Anthropocene (Crutzen 2002), which occupies much of the later Holocene.

We have lost some of the most potentially interesting species, many of them possibly with characteristics, including narrow habitat requirements, behaviors that made them susceptible to predation, and an inability to compensate for increased mortality through an increase in fecundity, that made them more vulnerable than the surviving species. What were the ecological and behavioral characteristics of ground sloths (megalonychids, megatheriids, nothrotheriids, mylodontids), the thylacine (*Thylacinus cynocephalus*), Cuban solenodon (*Solenodon cubanus*), Steller's sea cow (*Hydrodamalis gigas*), Labrador duck (*Camptorhynchus labradorius*), and ivory-billed (*Campephilus principalis*) and imperial (*C. imperialis*) woodpeckers that made them susceptible to human interference? Would our understanding of the biology of flightless birds change if we had aepyornis (*Aepyornis maximus*), moas (*Dinornis, Euryapteryx, Megalapteryx*), the Atitlán grebe (*Podilymbus gigas*), dodo (*Raphus cucullatus*), Rodrigues solitary (*Pezophaps solitaria*), and Stephens Island wren (*Xenicus lyalli*) available to study? Alan Tennyson relates the story that "[in] the 1980s a bird known as Kaua'ʻŌʻō (*Moho brac-catus*) often called [on the island of Kauaʻi]. Its call was never answered, for it was the last member of its species and the last of a remarkable genus of Hawaiian honeyeaters that was abundant and widespread before human settlement" (Tennyson & Martinson 2006).

A surplus of people

Unfortunately, the extinction of the fauna continues today. The increase in the human population will continue indefinitely until some biological limit is placed on it. Indeed, *there is already a surplus of people*. The present human world population is about 6.8 billion (U.S. Census Bureau 2009); it increases by approximately 203,800 each day and is estimated to reach 9.0 billion by 2040 (World Population Clock, International Data Base). Given these numbers, the extinction of species will continue. The African megafauna is disappearing; Canadians are clubbing 350,000 seals each year; Alaskans (with the help of a flaky ex-governor), Idahoans, and Coloradans are kill-ing wolves; the Utah legislature required the killing or removal of all wolves in the state; Japanese fake research promotes the consumption of dolphins and whales; overfishing of tunas is permitted to (temporarily) protect fish-eries; forests are being destroyed; air is being polluted; glaciers are melt-ing; fishnets and plastic are accumulating in the ocean; and the hunting of the formerly endangered sandhill crane (*Grus canadensis*) expands in the United States. *Homo sapiens*, which is not very sapient, has had all these effects while making a fetish of its own existence. A characteristic of many predators is that they do not tolerate competing predators, which is one reason why humans hunt bears, wolves, and large cats. *H. sapiens* is now the top predator. So, if Utah wants to get rid of its top predator, it should ask all people to leave.

Even if all people were ecologically sensitive—an absurd hope in a self-centered, greedy world—the growth of the human population would com-pletely overwhelm all species that need space and a quality environment to exist. The only possible hope for the nonhuman biological world would be to control—and reduce—the human population. However, even the mod-est goal of controlling the growth of this population is politically unaccept-able. The Koreans had days off from work to encourage people to go home and procreate! ("It is good for the economy.") The fundamental difficulties facing all conservation initiatives in the United States are the collective ig-norance of its citizens of the scientific basis underpinning all such proposals and the greed of people who make a profit by exploiting natural resources and dumping wastes.

Human impacts

Human activities, including the rampant cutting of forests, the dumping of wastes, including plastics (which will be a major contribution that we make to the fossil record), and the contamination of fresh and marine waters, have had many negative impacts on the condition of this planet. Of course, a major negative impact of humans, both direct and indirect, has been the extinction of elements of the native fauna. In New Zealand, for example, approximately 130 species of birds disappeared after the arrival of people; 27% were eliminated by humans and 70% by the predatory mammals they introduced (Tennyson & Martinson 2006). One of the most widespread human modifications of the environment is the modification of the chemical composition of the atmosphere, the principal consequence of which is global warming.

Global warming

According to the 2007 report of the Intergovernmental Panel on Climate Change (IPCC), the average global surface temperature increase in the twentieth century was 0.74% ± 0.18°C. This change was principally due to the accumulation of greenhouse gases in the atmosphere, predominantly carbon dioxide, but also including methane and nitrous oxide. The increase in these gases is a result of the burning of organic fuels (coal, wood, natural gas, oil) and the cutting of forests. Forests reduce atmospheric CO_2 by fixing CO_2 as organic compounds. Citizens of the United States constitute less than 5% of the world population, but produce approximately 25% of the CO_2 released into the atmosphere, an amount exceeded now only by China. As a result, eleven of the twelve warmest global surface temperatures since 1850 were recorded in the years from 1995 to 2005. Various projections by climatologists indicate that global surface temperatures in the twenty-first century will increase between 1.1°C and 6.4°C, depending on the climatic models used. The consequences of a further increase in surface temperatures will include the melting of polar ice caps and mountain glaciers, a change in the geographic pattern of precipitation with the resulting growth of subtropical deserts, and a rise in sea level with the disappearance of many oceanic islands. Global average sea level rose at an average of 1.8 (1.3–2.3) mm per year from 1961 to 2003, but at 3.1 (2.4–3.8) mm per year from 1993 to 2003. People who have a vested political or economic interest in supporting the status quo, of course, have denied these changes.

What are the implications of these changes for the distribution and survival of the native fauna? One of the clearest, which is now occurring, is the melting of the ice cap at the North Pole: since 1978, the average annual Arctic sea ice extent has shrunk by 2.7% (2.1%–3.3%) per decade. This shrinking has already had a severe impact on polar bears (*Ursus maritimus*) and walruses (*Odobenus rosmarus*), which depend on pack ice for reproduction and feeding; these species are now being forced to occupy tundra (bears) or land shorelines (walruses). If the pack ice were to disappear completely, these species would become extinct. This also applies to species at the South Pole: most penguins will be endangered as the southern ice cap melts, a trend that is already being seen in Adélie penguins (*Pygoscelis adeliae*) (Ainley 2002).

Physiological ecology is eminently qualified to suggest some of the consequences of global warming because, as we saw in chapter 12, aspects of the physiology of species are associated with their limits to distribution. Little can be done for species that live at the two poles if global warming continues: polar species cannot go farther north or south. Temperate species, however, may be able to move in a polar direction to maintain a distribution in their optimal climate (Hickling et al. 2006), as long as their movement is not precluded. Species that live in a topographically rich area may have the opportunity to move to higher altitudes (Tryjanowski et al. 2005; Hickling et al. 2006; Lenoir et al. 2008), unless the availability of water decreases due to a change in rainfall patterns, which may be important for plants (and their dependent animals) (Crimmins et al. 2010).

Global warming will have a great impact on the distribution of all terrestrial and aquatic species, depending on the local changes in weather. Southern areas of the temperate zone will become tropical—rainforests, if the rain is sufficient, or deserts, if rainfall is scarce. Thus, the desert Southwest of North America will expand, possibly including the Midwest. In any case, great changes in the distributions of terrestrial vertebrates will occur. Canada will get an array of terrestrial reptiles, including many snakes and lizards that will move far into its interior. Regions that today have tropical rainforests may well become xeric, and their endemic species will either move into warm, moist areas that now are in the temperate zone or will stay in place and go extinct. A further consequence of global warming for endotherms might be a change in body size (Gardner et al. 2011), as we saw in chapter 6.

So, what can concerned biologists do about this problem? Not much. A limited approach would be to isolate a few oceanic islands and restock them with flightless rails and other vulnerable island endemics (Steadman & Martin 2003). If this strategy is to be effective, people must be kept off these islands. As Steadman (2006) writes, "To see if flightlessness might evolve from currently volant populations of rails (Burney et al. 2002) would be interesting. Future biologists would be grateful if we began long-term genetic, morphological, and ecological monitoring of rail populations on islands uncontaminated with rats, cats, dogs, pigs, etc. Even if flightlessness did not evolve, the data would help us to understand the population biology of island rails, a most worthy cause." However, many small islands will be lost as a result of the rising sea level. (Imagine the political chaos resulting from the human movement away from island states and continental river valleys that this flooding will produce!)

What little else we can do is to study the living fauna exhaustively before it goes, just as Pliny the Elder summarized what was known in 77–79 in his *Naturalis Historia*, or as George-Louis Leclerc, Comte de Buffon, did in *Histoire naturelle* in 1749–1788. Such a program would catalogue the behavior and ecology of the fauna that is present today, an activity that will be appreciated by future generations of people who will live in a natural world more depleted than ours.

This concern becomes increasingly important in the face of the rapid increase in human population. Unfortunately, we human primates have developed advanced societies based on a commitment to exceedingly high energy expenditures, which we have extended from our bodies to our houses, cities, and work, with the resulting dependence on large deposits of coal, oil, and natural gas, in spite of the polluting nature of these energy sources. Furthermore, the attempt to bring all people up to a high standard of living would put an unsustainable burden on the planet. Wilson (2002) estimated that if all living people were to have the standard of living found today in Western Europe or North America, the resource base of four earthly planets would be required. Today, countries such as the People's Republic of China are buying land in poor countries to corner food production; these actions are likely to become a source of conflict in the future. Vital resources, such as high-quality water supplies, will be exhausted and will also be the basis of international conflict. And if that happens, human societies will probably collapse (see Wilson 2002; Weisman 2007).

Of course, the most effective solution to the looming overpopulation of humans is the extinction of *Homo sapiens*. Extinction ultimately happens to all species (where are the dinosaurs?), and it will happen to us as well, one way or another. Then the evolution of endothermic diversity can start again, possibly from domestic chickens, muscovy ducks, house mice, rats, cats, dogs, horses, and cattle. Maybe some other "intelligent" life will re-evolve out of the chaos, one that might be less destructive and self-important, a view suggested by Pete Scholander (1990). Maybe they will find in our artifacts that some of us tried to understand the world in which we lived. Maybe there are too many maybes.

REFERENCES

Aalto, M., A. Górecki, R. Meczeva, H. Wallgren & J. Weiner. 1993. Metabolic rates of the bank voles (*Clethrionomys glareolus*) in Europe along a latitudinal gradient from Lapland to Bulgaria. Annales Zoologici Fennici 30:233–238.

Ainley, D. G. 2002. The Adélie Penguin: Bellwether of Climate Change. Columbia University Press, New York.

Aitchison, C. W. 1987. Review of winter trophic relations of soricine shrews. Mammal Review 17:1–24.

Alder, G. H. & R. Levans. 1994. The island syndrome in rodent populations. Quarterly Review of Biology 69:473–490.

Aleksiuk, M. & I. M. Cowan. 1969a. Aspects of seasonal energy expenditure in the beaver (*Castor canadensis* Kuhl) at the northern limit of its distribution. Canadian Journal of Zoology 47:471–481.

———. 1969b. The winter metabolic depression in Arctic beavers (*Castor canadensis* Kuhl) with comparisons to California beavers. Canadian Journal of Zoology 47:965–979.

Alexander, R. L. 1995. Evidence of a counter-current heat exchanger in the ray, *Mobula tarapacana* (Chondrichthyes: Elasmobranchii: Batoidea: Myliobatiformes). Journal of Zoology, London 237:377–384.

———. 1996. Evidence of brain-warming in the mobulid rays, *Mobula tarapacana* and *Manta birostris* (Chondrichthyes: Elasmobranchii: Batoides: Myliobatiformes). Zoological Journal of the Linnean Society 118:151–164.

Ambrose, S. J. & S. D. Bradshaw. 1988. Seasonal changes in standard metabolic rates in the white-browed scrubwren *Sericornis frontalis* (Acanthizidae) from arid, semi-arid and mesic environments. Comparative Biochemistry and Physiology A 89:79–83.

Anderson, K. J. & W. Jetz. 2005. The broad-scale ecology of energy expenditure of endotherms. Ecology Letters 8:310–318.

Anderson, M. D., J. B. Williams & P. R. K. Richardson. 1997. Laboratory metabolism and evaporative water loss of the aardwolf, *Proteles cristatus*. Physiological Zoology 70:464–469.

Andrewartha, H. G. & L. C. Birch. 1954. The Distribution and Abundance of Animals. University of Chicago Press, Chicago.

Appenzeller, T. 1994. Paleontology: Argentine dinos vie for heavyweight title. Science 266:1805.

Archer, M., T. F. Flannery, A. Richie & R. E. Molnar. 1985. 1st Mesozoic mammal from Australia: An Early Cretaceous monotreme. Nature 318:363–366.

Arends, A. & B. K. McNab. 2001. The comparative energetics of "caviomorph" rodents. Comparative Biochemistry and Physiology A 130:105–122.

Arieli, R. 1990. Adaptation of the mammalian gas transport system to subterranean life. In Evolution of Subterranean Mammals at the Organismal and Molecular Levels, edited by E. Nevo & O. A. Reig, 251–268. Wiley-Liss, New York.

Arkins, A. M., A. P. Winnington & S. Anderson. 1999. Diet and nectarivorous foraging behaviour of the short-tailed bat (*Mystacina tuberculata*). Journal of Zoology, London 247:183–187.

Arlettaz, R., C. Ruchet, J. Aeschimann, E. Brun, M. Genoud & P. Vogel. 2000. Physiological traits affecting the distribution and winter strategy of the bat *Tadarida teniotis*. Ecology 81:1004–1014.

Aschoff, J. 1981. Thermal conductance in mammals and birds: its dependence on body size and circadian phase. Comparative Biochemistry and Physiology A 69:611–619.

Aschoff, J. & H. Pohl. 1970a. Der Ruheumsatz von Vögeln als Funktion der Tageszeit und der Körpergröße. Journal für Ornithologie 111:38–47.

———. 1970b. Rhythmic variations in energy metabolism. Federation Proceedings 29:1541–1552.

Ashton, K. G. 2002. Patterns of within-species body size variation of birds: Strong evidence for Bergmann's rule. Global Ecology and Biogeography 11:505–523.

Ashton, K. G., M. C. Tracy & A. de Queiroz. 2000. Is Bergmann's rule valid for mammals? American Naturalist 156:390–415.

Auffenberg, W. 1981. The Behavioral Ecology of the Komodo Monitor. University Presses of Florida, Gainesville.

Bakker, R. T. 1975. Experimental and fossil evidence for the evolution of tetrapod bioenergetics. In Perspectives in Biophysical Ecology, edited by D. M. Gates & R. Schmerl, 365–399. Springer-Verlag, New York.

———. 1980. Dinosaur heresy-dinosaur renaissance. In A Cold Look at Warm-Blooded Dinosaurs, edited by R. D. K. Thomas & E. C. Olson, 351–462. Westview Press, Boulder, Colorado.

———. 1986. The Dinosaur Heresies. William Morrow and Company, New York.

Balouet, J. C. 1991. The fossil vertebrate record of New Caledonia. In Vertebrate Palaeontology of Australasia, edited by P. Vickers-Rich, J. M. Monaghan, R. F. Baird & T. H. Rich, 1383–1409. Monash University, Melbourne, Australia.

Balouet, J. C. & E. Buffetaut. 1987. *Mekosuchus inexpectatus*, n. g., n. sp., crocodilien nouveau de l'Holocene de Nouvelle Caledonie. Compte Rendue de l'Académie des Sciences Paris (Série 2) 304:853–856.

Banaver, J. R., J. Damuth, A. Maritan & A. Rinaldo. 2002. Supply-demand balance and metabolic scaling. Proceedings of the National Academy of Sciences, USA 99:10506–10509.

Banaver, J. R., A. Maritan & A. Rinaldo. 1999. Size and form in efficient transportation networks. Nature 399:130–132.

Banfield, A. W. F. 1974. The Mammals of Canada. University of Toronto Press, Toronto.

Baptista, L. F., P. W. Trail & H. M. Horblit. 1997. Family Columbidae (Pigeons and Doves). In Handbook of the Birds of the World, vol. 4, Sandgrouse to Cuckoos, edited by J. del Hoyo, A. Elliott & J. Sargatal, 60–243. Lynx Editions, Barcelona.

Barker, F. K., A. Cibois, P. A. Schikler, J. Feinstein & J. Cracraft. 2004. Phylogeny and diversification of the largest avian radiation. Proceedings of the National Academy of Sciences, USA 101:11040–11045.

Barrett, I. & F. J. Hester. 1964. Body temperature of yellowfin and skipjack tunas in relation to sea surface temperature. Nature 203:96–97.

Bartholomew, G. A. 1977. Body temperature and energy metabolism. In Animal Physiology: Principles and Adaptations, 3rd ed., edited by M. S. Gordon, 364–449. MacMillan, New York.

Bartholomew, G. A. & T. J. Cade. 1957. Temperature regulation, hibernation, and aestivation in the little pocket mouse, *Perognathus longimembris*. Journal of Mammalogy 38:60–72.

———. 1963. The water economy of land birds. Auk 80:504–539.

Bartholomew, G. A. & J. W. Hudson. 1960. Aestivation in the Mohave ground squirrel, *Citellus mohavensis*. Bulletin of the Museum of Comparative Zoology 124:193–208.

Bartholomew, G. W. & C. H. Trost. 1970. Temperature regulation in the speckled mousebird, *Colius striatus*. Condor 72:141–146.

Bartholomew, G. W., C. M. Vleck & T. L. Bucher. 1983. Energy metabolism and nocturnal hypothermia in two tropical passerine frugivores, *Manacus vitellinus* and *Pipra mentalis*. Physiological Zoology 56:370–379.

Beauchamp, A. J. 1989. Panbiogeography and rails of the genus *Gallirallus*. New Zealand Journal of Zoology 16:763–772.

Benedict, F. G. 1938. Vital Energetics: A Study of Comparative Basal Metabolism. Publication 425. Carnegie Institute, Washington, DC.

Bennett, A. F. & W. R. Dawson. 1976. Metabolism. In Biology of the Reptilia, vol. 5, edited by C. Gans & W. R. Dawson, 127–223. Academic Press, New York.

Bennett, A. F., J. W. Hicks & A. J. Cullum. 2000. An experimental test of the thermoregulatory hypothesis for the evolution of endothermy. Evolution 54:1768–1773.

Bergmann, C. 1847. Ueber die Vehältnisse der Wärmeökonomie der Tiere zu ihrer Grösse. Göttinger Studien 3:595–708.

Bertelli, S., L. M. Chiappe & C. Tambussi. 2007. A new phorusrhacid (Avis: Cariamae) from the Middle Miocene of Patagonia, Argentina. Journal of Vertebrate Paleontology 27:409–419.

Beuchat, C. A., S. B. Chaplin & M. L. Morton. 1979. Ambient temperature and daily energetics of two species of hummingbirds, *Calypte anna* and *Selasphorus rufus*. Physiological Zoology 52:280–295.

Birt-Friesen, V. L., W. A. Montevecchi, D. K. Cairns & S. A. Macko. 1989. Activity specific metabolic rates of free-living northern gannets and other seabirds. Ecology 70:357–367.

Block, B. A. 1986. Structure of the brain and eye heater tissue in marlins, sailfish, and spearfishes. Journal of Morphology 190:169–189.

Block, B. A. & F. G. Carey. 1985. Warm brain and eye temperatures in sharks. Journal of Comparative Physiology B 156:229–236.

Block, B. A. & J. R. Finnerty. 1994. Endothermy in fishes: A phylogenetic analysis of constraints, predispositions, and selection pressures. Environmental Biology of Fishes 40:283–302.

Block, B. A., J. R. Finnerty, A. F. R. Stewart & J. Kidd. 1993. Evolution of endothermy in fish: Mapping physiological traits on a molecular phylogeny. Science 260:210–214.

Boily, P. & D. M. Lavigne. 1995. Resting metabolic rates and respiratory quotients of gray seals (*Halichoerus grypus*) in relation to time of day and duration of food deprivation. Physiological Zoology 68:1181–1193.

———. 1997. Developmental and seasonal changes in resting metabolic rates of captive female grey seals. Canadian Journal of Zoology 75:1781–1789.

Bokma, F. 2004. Evidence against universal metabolic allometry. Functional Ecology 18:184–187.

Bonaccorso, F. J. & B. K. McNab. 1997. Plasticity of energetics in blossom bats (Pteropodidae): Impact on distribution. Journal of Mammalogy 78:1073–1088.

———. 2003. Standard energetics of leaf-nosed bats (Hipposideridae): Its relationship to intermittent- and protracted-foraging tactics in bats and birds. Journal of Comparative Physiology B 173:43–53.

Bothma, J. du P. & J. A. J. Nel. 1980. Winter food and foraging behaviour of the aardwolf *Proteles cristatus* in the Namib-Naukluft Park. Madoqua 12:141–145.

Bourn, D. & M. Coe. 1978. Size, structure and distribution of giant tortoise population of Aldabra. Philosophical Transactions of the Royal Society of London B 282:139–175.

Bowers, J. R. 1971. Resting metabolic rate in the cotton rat *Sigmodon*. Physiological Zoology 44:137–148.

Boyce, M. S. 1978. Climatic variability and body size variation in the muskrats (*Ondatra zibethicus*) of North America. Oecologia 36:1–19.

———. 1979. Seasonality and patterns of natural selection for life histories. American Naturalist 114:569–583.

Boyd, I. L. 2002. Energetics: Consequences for fitness. In Marine Mammal Biology: An Evolutionary Approach, edited by A. R. Hoelzel, 247–277. Blackwell Science, Oxford.

Bozinovic, F., R. Gricelda & M. Rosenmann. 2004. Energetics and torpor of a South American "living fossil," the microbiotheriid *Dromiciops gliroides*. Journal of Comparative Physiology B 174:293–297.

Bozinovic, F., J. M. Rojas, B. R. Broitman & R. A. Vásquez. 2009. Basal metabolism is correlated with habitat productivity among populations of degus (*Ocotodon degus*). Comparative Biochemistry and Physiology A 152:560–564.

Bozinovic, F., G. Ruiz, A. Cortés & M. Rosenmann. 2005. Energetics, thermoregulation and torpor in the Chilean mouse-opossum *Thylamys elegans* (Didelphidae). Revista Chilena de Historia Natural 78:199–206.

Bradley, S. R. & D. R. Deavers. 1980. A reexamination of the relationship between thermal conductance and body weight in mammals. Comparative Biochemistry and Physiology A 65:465–476.

Bradley, S. R. & J. W. Hudson. 1974. Temperature regulation in the tree shrew *Tupaia glis*. Comparative Biochemistry and Physiology A 48:55–60.

Bradley, W. G., J. S. Miller & M. K. Yousef. 1974. Thermoregulatory patterns in pocket gophers: Desert and mountain. Physiological Zoology 47:172–179.

Bradley, W. G. & M. K. Yousef. 1975. Thermoregulatory responses in the plains pocket gopher, *Geomys bursarius*. Comparative Biochemistry and Physiology A 52:35–38.

Brigham, R. M. 1992. Daily torpor in a free-ranging goatsucker, the common poorwill (*Phalaenoptilus nuttallii*). Physiological Zoology 65:457–472.

Brigham, R. M., G. Körtner, T. A. Maddocks & F. Geiser. 2000. Seasonal use of torpor by free-ranging Australian owlet-nightjars (*Aegotheles cristatus*). Physiological and Biochemical Zoology 73:613–620.

Brocke, R. H. 1970. The winter ecology and bioenergetics of the opossum, *Didelphis marsupialis*, as distributional factors in Michigan. Ph.D. thesis, Michigan State University, East Lansing, Michigan.

Brown, C. R., E. M. Hunter & R. M. Baxter. 1997. Metabolism and thermoregulation in the forest shrew *Myosorex varius* (Soricidae: Crocidurinae). Comparative Biochemistry and Physiology A 118:1285–1290.

Brown, R. P. & V. Perez-Mellado. 1994. Ecological energetics and food acquisition in dense Menorcan islet populations of the lizard *Podarcis lilifordi*. Functional Ecology 8:427–434.

Bryant, D. M. & R. W. Furness. 1995. Basal metabolic rates of North-Atlantic seabirds. Ibis 137:219–226.

Buffenstein, R. & J. U. M. Jarvis. 1985. Thermoregulation and metabolism in the smallest African gerbil, *Gerbillus pusillus*. Journal of Zoology, London A 205:107–121.

Buffenstein, R. & S. Yahav. 1991. Is the naked mole-rat *Heterocephalus glaber* an endothermic yet poikilothermic mammal? Journal of Thermal Biology 16:227–232.

Buffetaut, E. 1983. Sur la persistence tardive d'un crocodilien archaïque dans le Pleistocene d'Ile des Pins (Nouvelle-Calédonie) et sa signification biogeographique. Compte Rendus de l'Académie des Sciences Paris (Série 2) 297:89–92.

Bullen, R. D. & N. L. MacKenzie. 2004. Bat flight muscle mass: Implications for foraging strategy. Australian Journal of Zoology 52:605–622.

Burda, H. 2006. Ear and eye in subterranean mole-rats, *Fukomys anselli* (Bathyergidae) and *Spalax ehrenbergi* (Spalacidae): Progressive specialization or regressive degeneration? Animal Biology 56:475–486.

Burney, D. A., D. W. Steadman & P. S. Martin. 2002. Evolution's second chance. Wild Earth 12:12–15.

Burton, S., M. R. Perrin & C. T. Downs. 2008. The thermal biology of African lovebirds and grass parakeets. Journal of Thermal Biology 33:355–362.

Butler, P. J. & A. J. Woakes. 2001. Seasonal hypothermia in a large migrating bird: Saving energy or fat deposition. Journal of Experimental Biology 204:1361–1367.

Calder, W. A. 1984. Size, Function, and Life History. Harvard University Press, Cambridge, Massachusetts.

Cameron, A. W. 1958. Mammals of the islands in the Gulf of St. Lawrence. Bulletin of the National Museum of Canada 154:1–165.

Campbell, K. E. & E. P. Tonni. 1983. Size and locomotion in teratorns (Aves, Teratornithidae). Auk 100:390–403.

Canterbury, G. 2002. Metabolic adaptation and climatic constraints on winter bird distribution. Ecology 83:946–957.

Careau, V., J. Morand-Ferron & D. Thomas. 2007. Basal metabolic rate of Canidae from hot deserts to cold arctic climates. Journal of Mammalogy 88:394–400.

Carey, F. G. 1982. A brain heater in the swordfish. Science 216:1327–1329.

Carey, F. G., G. Gabrielson, J. W. Kanwisher, O. Brazier, J. G. Casey & H. L. Pratt, Jr. 1982. The white shark, *Carcharodon carcharias*, is warm-bodied. Copeia 1982:254–260.

Carey, F. G. & J. M. Teal. 1969. Mako and porbeagle: Warm-bodied sharks. Comparative Biochemistry and Physiology 28:199–204.

Carey, F. G., J. M. Teal & J. W. Kanwisher. 1981. The visceral temperatures of mackerel sharks (Lamnidae). Physiological Zoology 54:334–344.

Carey, F. G., J. M. Teal, J. W. Kanwisher, K. D. Lawson & J. S. Beckett. 1971. Warm-bodied fish. American Zoologist 11:137–145.

Carpenter, F. L. & M. A. Hixon. 1988. A new function for torpor: Fat conservation in a wild migrant hummingbird. Condor 90:373–378.

Carpenter, R. E. 1966. A comparison of thermoregulation and water metabolism in the kangaroo rats *Dipodomys agilis* and *Dipodomys merriami*. University of California Publications in Zoology. University of California Press, Berkeley.

Castro, G. 1989. Energy costs and avian distributions: Limitations or chance?—A comment. Ecology 70:1181–1182.

Cavieres, G. & P. Sabat. 2008. Geographic variation in the response to thermal acclimation in rufous-collard sparrows: Are physiological flexibility and environmental heterogeneity correlated? Functional Ecology 22:509–515.

Chaffee, R. R. J. & J. C. Roberts. 1971. Temperature acclimation in birds and mammals. Annual Review of Physiology 33:155–202.

Chappell, M. A., V. H. Shoemaker & D. N. Janes. 1993. Energetics of foraging in breeding Adelie penguins. Ecology 74:2450–2461.

Chevalier, C. D. 1987. Comparative thermoregulation in tropical procyonids. American Zoologist 27:A146.

Chiappe, L. M. & S. Bertelli. 2006. Skull morphology of giant terror birds. Nature 443:929.

Churchfield, S., V. A. Nesterenko & E. A. Shvarts. 1999. Food niche overlap and ecological separation amongst six species of coexisting forest shrews (Insectivora: Soricidae) in the Russian Far East. Journal of Zoology, London 248:349–359.

Clarke, A., P. Rothery & N. J. B. Issac. 2010. Scaling of basal metabolic rate with body mass and temperature in mammals. Journal of Animal Ecology 79:616–619.

Cockrum, E. L. 1969. Migration in the guano bat, *Tadarida brasiliensis*. In Contributions in Mammalogy, edited by J. K. Jones, Jr., 303–336. Miscellaneous Publications of the Museum of Natural History, University of Kansas 51. University of Kansas, Lawrence.

Cody, M. L. 1966. A general theory of clutch size. Evolution 20:174–184.

Collar, N. J. 2005. Turdidae (Thrushes). In Handbook of the Birds of the World, vol. 10, Cuckoo-Shrikes to Thrushes, edited by J. del Hoyo, A. Elliott & D. Christie, 514–807. Lynx Editions, Barcelona.

Contreras, L. C. 1986. Bioenergetics and distribution of fossorial *Spalacopus cyanus* (Rodentia): Thermal stress, or cost of burrowing? Physiological Zoology 59:20–28.

Contreras, L. C. & B. K. McNab. 1990. Thermoregulation and energetics of subterranean mammals. In Evolution of Subterranean Mammals at the Organismal and Molecular Levels, edited by E. Nevo & O. A. Reig, 231–250. Wiley-Liss, New York.

Cooper, C. E. & F. Geiser. 2008. The "minimum boundary curve for endothermy" as a predictor of heterothermy in mammals and birds: A review. Journal of Comparative Physiology B 178:1–8.

Cooper, C. E. & P. C. Withers. 2006. Numbats and aardwolves—how low is low? A reaffirmation of the need for statistical rigour in evaluating regression predictions. Journal of Comparative Physiology B 176:623–629.

Cooper, R. L. & J. D. Skinner. 1979. Importance of termites in the diet of the aardwolf *Proteles cristatus* in South Africa. South African Journal of Zoology 14:5–8.

Corbett, L. 1995. The Dingo in Australia and Asia. Comstock/Cornell University Press, New York.

Corp, N., M. L. Gorman & J. R. Speakman. 1999. Daily energy expenditure of free-living male wood mice in different habitats and seasons. Functional Ecology 13:585–593.

Costa, D. P. 2001. Energetics. In Encyclopedia of Marine Mammals, edited by W. F. Perrin, J. G. M. Thewissen & B. Wursig, 387–394. Academic Press, New York.

Costa, D. P., J. P. Croxall & C. Duck. 1989. Foraging energetics of Antarctic fur seals, *Arctocephalus gazella*, in relation to changes in prey availability. Ecology 70:596–606.

Costa, D. P. & P. A. Prince. 1987. Foraging energetics of gray-headed albatrosses *Diomedea chrysostoma* at Bird Island, South Georgia. Ibis 129:149–158.

Costa, D. P. & T. M. Williams. 1999. Marine mammal energetics. In The Biology of Marine Mammals, edited by J. Reynolds & T. Twiss, 176–217. Smithsonian Institution Press, Washington, DC.

Cracraft, J. & J. Feinstein. 2000. What is not a bird of paradise? Molecular and morphological evidence places *Macgregoria* in the Meliphagidae and the Cnemopilinae near the base of the corvoid tree. Proceedings of the Royal Society of London B 267:322–241.

Crimmins, S. M., S. Z. Dobrowski, J. A. Greenberg, J. T. Abatzoglou & A. R. Mynsberge. 2010. Changes in climatic water balance drive downhill shifts in plant species' optimum elevations. Science 331:324–327.

Crompton, A. W., C. R. Taylor & J. A. Jagger. 1978. Evolution of homeothermy in mammals. Nature 272:333–336.

Crutzen, P. J. 2002. Geology of mankind. Nature 415:23.

Cruz-Neto, A. P. & F. Bozinovic. 2004. The relationship between diet quality and basal metabolic rate in endotherms: Insights from intraspecific analysis. Physiological and Biochemical Zoology 77:877–889.

Cruz-Neto, A. P., T. Garland, Jr. & A. S. Abe. 2001. Diet, phylogeny, and basal metabolic rate in phyllostomid bats. Zoology 104:49–58.

Daan, S., D. Masman & A. Groenewold. 1990. Avian basal metabolic rates: Their association with body composition and energy expenditure in nature. American Journal of Physiology 259:R333–R340.

Daniel, M. J. 1979. The New Zealand short-tailed bat, *Mystacina tuberculata*; a review of present knowledge. New Zealand Journal of Zoology 6:357–370.

Daniels, H. L. 1984. Oxygen consumption in *Lemur fulvus*: Deviation from the ideal model. Journal of Mammalogy 65:584–592.

Dausmann, K. H., J. Glos, J. U. Ganzhorn & G. Heldmaier. 2004. Hibernation in a tropical primate. Nature 429:825–826.

Davis, W. H. 1965. Winter awakening patterns in the bat *Myotis lucifugus* and *Pipistrellus subfarus*. Journal of Mammalogy 45:645–647.

Dawson, T. J. & J. M. Olson. 1987. The summit metabolism of the short-tailed shrew *Blarina brevicauda*: A high summit is further elevated by cold acclimation. Physiological Zoology 60:631–639.

Dawson, W. R. & A. F. Bennett. 1973. Roles of metabolic level and temperature regulation in adjustment of western plumed pigeons (*Lophophaps ferruginea*) to desert conditions. Comparative Biochemistry and Physiology A 44:249–266.

Dawson, W. R., A. F. Bennett & J. W. Hudson. 1976. Metabolism and thermoregulation in hatchling ring-billed gulls. Condor 78:49–60.

Dawson, W. R. & C. D. Fisher. 1969. Responses to temperature by spotted nightjar (*Eurostopodus guttatus*). Condor 71:49–53.

Dawson, W. R. & G. C. Whittow. 2000. Regulation of body temperature. In Sturkie's Avian Physiology, 5th ed., edited by G. C. Whittow, pp. 343–390. Academic Press, San Diego, California.

Dayan, T., D. Simberloff, E. Tchernov & Y. Yom-Tov. 1991. Calibrating the paleothermometer: Climate, communities, and the evolution of body size. Paleobiology 17:189–199.

Deavers, D. R. & J. W. Hudson. 1981. Temperature regulation in 2 rodents (*Clethrionomys* gapperi and *Peromyscus leucopus*) and a shrew (*Blarina brevicauda*) inhabiting the same environment. Physiological Zoology 54:94–108.

Dehnel, A. 1949. Studies on the genus *Sorex* L. Annales of Maria Curie-Sklodowska University Section C Biology 4:17–102.

de Muizon, C. & H. G. McDonald. 1995. An aquatic sloth from the Pliocene of Peru. Nature 375:224–227.

de Muizon, C., H. G. McDonald, R. Salas & M. Urbina. 2004. The youngest species of the aquatic sloth *Thalassocnus* and a reassessment of the relationships of the nothrothere sloths (Mammalia: Xenarthra). Journal of Vertebrate Paleontology 24:387–397.

de Ricqlès, A. J. 1974. Evolution of endothermy: Histological evidence. Evolutionary Theory 1:51–80.

Dizon, A. E. & R. W. Brill. 1979. Thermoregulation in tunas. American Zoologist 19:249–265.

Döbler, H.-J. 1982. Temperaturregulation und Saurstoffverbrauch beim Senegal- und Zwerggalago [*Galago senegalensis, Galago (Galagoides) demidovii*]. Bonner zoologische Beiträge 33:33–59.

Domning, D. P. 1978. Sirenian Evolution in the North Pacific Ocean. University of California Publications in Geological Sciences 118. University of California Press, Berkeley.

Dondini, G. & S. Vergari. 2000. Carnivory in the greater noctule bat (*Nyctalus lasiopterus*) in Italy. Journal of Zoology, London 251:233–236.

Downs, C. T. & M. Brown. 2002. Nocturnal heterothermy and torpor in the malachite sunbird (*Nectarinia famosa*). Auk 119:251–260.

Downs, C. T. & M. R. Perrin. 1996. The thermal biology of southern Africa's smallest rodent, *Mus minutoides*. South African Journal of Science 92:282–285.

Drent, R. H. & M. Klaassen. 1989. Energetics of avian growth: The causal link with BMR and metabolic scope. In Physiology of Cold Adaptation in Birds, edited by C. Bech & R. E. Reinertsen, 349–359. Plenum, New York.

Driskell, A., L. Christidis, B. J. Gill, W. E. Boles, F. K. Barker & N. W. Longmore. 2007. A new endemic family of New Zealand passerine birds: Adding heat to a biodiversity hotspot. Australian Journal of Zoology 55:73–78.

Du Bois, E. F. 1936. Basal metabolism in health and disease. 3rd ed. Lea & Febiger, Philadelphia.

Dunbrack, R. L. & M. A. Ramsay. 1993. The allometry of mammalian adaptations to seasonal environments: A critique of the fasting endurance hypothesis. Oikos 66:336–342.

Dyck, A. P. & R. A. MacArthur. 1992. Seasonal patterns of body temperature and activity in free-ranging beaver (*Castor canadensis*). Canadian Journal of Zoology 70:1668–1672.

———. 1993. Daily energy requirements of beaver (*Castor canadensis*) in a simulated winter microhabitat. Canadian Journal of Zoology 71:2131–2135.

Ehlers, R. & M. L. Morton. 1982. Metabolic rate and evaporative water loss in the least seed-snipe, *Thinocorus rumicivorus*. Comparative Biochemistry and Physiology A 73:233–235.

Ehrhardt, N., G. Heldmaier & C. Exner. 2005. Adaptive mechanisms during food restriction in *Acomys russatus*: The use of torpor for desert survival. Journal of Comparative Physiology B 175:193–200.

Ellis, H. I. 1985. Energetics of free-ranging sea birds. In Seabird Energetics, edited by G. C. Whittow & H. Rahn, 203–234. Plenum, New York.

Enger, P. S. 1957. Heat regulation and metabolism in some tropical mammals and birds. Acta Physiologica Scandinavica 40:161–166.

Ericson, P. G. P., C. L. Anderson, T. Britton, A. Elzanowiski, U. S. Johansson, M. Källersjö, J. I. Ohlson, T. J. Parsons, D. Zuccon & G. Mayr. 2006. Diversification of Neoaves: Integration of molecular sequence data and fossils. Biology Letters 2:543–547.

Erlinge, S. 1987. Why do European stoats *Mustela erminea* not follow Bergmann's rule? Holarctic Ecology 10:33–39.

Fain, M. G. & P. Houde. 2004. Parallel radiations in the primary clades of birds. Evolution 58:2558–2573.

Falow, J. O., P. Dobson & A. Chinsamy. 1995. Dinosaur biology. Annual Review of Ecology and Systematics 26:445–471.

Feduccia, A. 1996. The Origin and Evolution of Birds. Yale University Press, New Haven.

Feduccia, A., T. Lingham-Soliar & J. R. Hinchliffe. 2005. Do feathered dinosaurs exist? Testing the hypothesis on neontological and paleontological evidence. Journal of Morphology 266:125–166.

Felsenstein, J. Phylogenies and the comparative method. American Naturalist 125:1–15.

Firman, M. C., R. M. Brigham & R. M. R. Barclay. 1993. Do free-ranging common nighthawks enter torpor? Condor 95:157–162.

Flannery, T. 1995. Mammals of New Guinea. Comstock/Cornell University Press, Ithaca, New York.

Flannery, T. F., M. Archer, T. H. Rich & R. Jones. 1995. A new family of monotremes from the Cretaceous of Australia. Nature 377:418–420.

Flannery, T. F. & C. P. Groves. 1998. A revision of the genus *Zaglossus* (Monotremata, Tachyglossidae), with description of new species and subspecies. Mammalia 62:367–396.

Fleischer, R. C. & R. F. Johnston. 1982. Natural selection on body size and proportions in house sparrows. Nature 298:747–749.

Fleming, T. H. 1986. Opportunism versus specialization: The evolution of feeding strategies in frugivorous bats. In Frugivores and Seed Dispersal, edited by A. Estrada & T. H. Fleming, 105–118. W. Junk, Dordrecht.

Folch, A. 1992. Family Apterygidae (kiwis). In Handbook of the Birds of the World, vol. 1, Ostrich to Ducks, edited by J. del Hoyo, A. Elliott & J. Sargatal, 104–100. Lynx Editions, Barcelona.

Fortelius, M. & J. Kappelman. 1993. The largest land mammal ever imagined. Zoological Journal of the Linnean Society 108:85–101.

Foster, J. B. 1964. Evolution of mammals on islands. Nature 202:234–235.

Frair, W., R. G. Ackman & N. Mrosovsky. 1972. Body temperature of *Dermochelys coriacea*: Warm turtle from cold water. Science 177:791–793.

Franklin, D. D. & K. J. Wilson. 2003. Are low reproductive rates characteristic of New Zealand's native terrestrial birds? Evidence from the allometry of nesting parameters in altricial species. New Zealand Journal of Zoology 30:185–204.

French, A. R. 1976. Selection of high temperatures for hibernation by the pocket mouse, *Perognathus longimembris*: Ecological advantages and energetic consequences. Ecology 57:185–191.

———. 1985. Allometries of the durations of torpid and euthermic intervals during mammalian hibernation: Test of the theory of metabolic control of the timing of changes in body temperature. Journal of Comparative Physiology B 156:13–19.

Frey, H. 1980. Le métabolisme énergétique de *Suncus etruscus* (Soricidae, Insectivora) en torpeur. Revue suisse de Zoologie 87:739–748.

Gaetner, R. A., J. S. Hart & O. Z. Roy. 1973. Seasonal spontaneous torpor in the white-footed mouse, *Peromyscus leucopus*. Comparative Biochemistry and Physiology A 45:169–181.

Gallivan, G. J. & R. C. Best. 1980. Metabolism and respiration of the Amazonian manatee (*Trichechus inunguis*). Physiological Zoology 53:245–253.

Gardner, J. L., A. Peters, M. R. Kearney, L. Joseph & R. Heinshon. 2011. Declining body size: A third universal response to warming? Trends in Ecology and Evolution 26:285–291.

Gates, D. M. 1962. Energy Exchange in the Biosphere. Harper & Row, New York.

———. Biophysical Ecology. Springer-Verlag, New York.

Gavrilov, V. M. 1995. The maximum, potential productive and normal levels of the metabolism of existence in passerine and non-passerine birds. 2. Correlations with the level of external work, energetics and ecological conditions. Zoological Zhurnal 74:108–123.

———. 1998. Rationale for the evolutionary increase in basal metabolic rate of Passeriformes. Doklady Biological Sciences 358:85–88.

———. 1999. Ecological phenomena of Passeriformes as a derivative of their energetics. Acta Ornithologica 34:165–172.

Gebczynski, M. & E. Szuma. 1993. Metabolic rate in *Pitymys subterraneus* of two coat-colour morphs. Acta Theriologica 38:291–296.

Geist, V. 1987. Bergmann's rule is invalid. Canadian Journal of Zoology 65:1035–1038.

Genoud, M. 1988. Energetic strategies of shrews: Ecological constraints and evolutionary implications. Mammal Review 18:173–193.

———. 1993. Temperature regulation in subtropical tree bats. Comparative Biochemistry and Physiology A 104:321–331.

Genoud, M. & M. Ruedi. 1996. Rate of metabolism, temperature regulations, and evaporative water loss in the lesser gymnure *Hylomys suillus* (Insectivora, Mammalia). Journal of Zoology, London 240:309–316.

Genoud, M. & P. Vogel. 1990. Energy requirements during reproduction and reproductive effort in shrews (Soricidae). Journal of Zoology, London 220:41–60.

Gettinger, R. D. 1975. Metabolism and thermoregulation of a fossorial rodent, the northern pocket gopher (*Thomomys talpoides*). Physiological Zoology 48:311–322.

Gillooly, J. F., A. P. Allen & E. L. Charnov. 2007. Dinosaur fossils predict body temperatures. PLoS Biology 4:1467–1469.

Gillooly, J. F., J. H. Brown, G. B. West, V. M. Savage & E. L. Charnov. 2001. Effects of size and temperature on metabolic rate. Science 293:2248–2251.

Glass, B. P. 1982. Seasonal movements of the Mexican freetail bats *Tadarida brasiliensis mexicana* banded in the Great Plains. Southwestern Naturalist 27:127–133.

Glazier, D. S. 1985a. Energetics of litter size in five species of *Peromyscus* with generalizations for other mammals. Journal of Mammalogy 66:629–642.

———. 1985b. Relationship between metabolic rate and energy expenditure for lactation in *Peromyscus*. Comparative Biochemistry and Physiology A 80:587–590.

———. 2005. Beyond the "$\frac{3}{4}$-power law": Variation in the intra- and interspecific scaling of metabolic rate in animals. Biological Reviews 80:611–662.

———. 2008. Effects of metabolic level on the body size scaling of metabolic rate in birds and mammals. Proceedings of the Royal Society of London B 278:1405–1410.

———. 2010. A unifying explanation for diverse metabolic scaling in animals and plants. Biological Reviews 85:111–138.

Goltsman, M., E. P. Kruchenkova, S. Sergeev, I. Volodin & D. W. Macdonald. 2005. The "island syndrome" in a population of Arctic foxes (*Alopex lagopus*) from Mednyi Island. Journal of Zoology, London 267:405–418.

Gorecki, A. & L. Christov. 1969. Metabolic rate of the lesser mole rat. Acta Theriologica 14:441–448.

Grant, P. R. 1972. Interspecific competition among rodents. Annual Review of Ecology and Systematics 3:79–106.

Grant, T. I. 1978. Adaptations of tissue and limb segments to facilitate moving and feeding in arboreal folivores. In The Ecology of Arboreal Folivores, edited by G. G. Montgomery, 231–241. National Zoological Park, Washington, DC.

Greegor, D. H., Jr. 1980. Diet of the little hairy armadillo, *Chaetophractus vellerosus*, of northwestern Argentina. Journal of Mammalogy 61:331–334.

Green, B., D. King, M. Braysher & A. Saim. 1991. Thermoregulation, water turnover, and energetics of free-living komodo dragons, *Varanus komodoensis*. Comparative Biochemistry and Physiology A 99:97–102.

Grodzinski, W., H. Böckler & G. Heldmaier. 1988. Basal and cold-induced metabolic rates in the harvest mouse *Micromys minutus*. Acta Theriologica 33:283–291.

Guilday, J. E. 1958. The prehistoric distribution of the opossum. Journal of Mammalogy 39:39–43.

Guthrie, R. D. 1984. Mosaics, allelochemicals and nutrients: An ecological theory of late Pleistocene megafaunal extinctions. In Quaternary Extinctions: A Prehistoric Revolution, edited by P. S. Martin & S. Klein, 254–298. University of Arizona Press, Tucson.

Hackett, S. J., R. T. Kimball, S. Reddy, R. C. K. Bowie, E. L. Braun, M. J. Braun, J. L. Chojnowski, W. A. Cox, K.-L. Han, J. Harshman, C. J. Huddleston, B. D. Marks, K. J. Migla, W. S. Moore, F. H. Sheldon, D. W. Steadman, C. C. Witt & T. Yuri. 2008. A phylogenomic study of birds reveals their evolutionary history. Science 320:1763–1768.

Haim, A. 1987a. Metabolism and thermoregulation in rodents: Are these adaptations to habitat and food quality? Suid-Afrikaanse Tydskrif vir Wetenskap 83:639–642.

———. 1987b. Thermoregulation and metabolism of Wagner's gerbil (*Gerbillus dasyurus*): A rock dwelling rodent adapted to arid and mesic environments. Journal of Thermal Biology 12:45–48.

Haim, A. & A. Borut. 1986. Reduced heat production in the bushy-tailed gerbil *Skeetamys calurus* (Rodentia) as an adaptation to arid environments. Mammalia 50:27–33.

Haim, A. & I. Izaki. 1993. The ecological significance of resting metabolic rate and non-shivering thermogenesis for rodents. Journal of Thermal Biology 18:71–81.

———. 1995. Comparative physiology of thermoregulation in rodents: Adaptations to arid and mesic environments. Journal of Arid Environments 31:431–440.

Haim, A., R. J. Van Aarde & J. D. Skinner. 1990. Metabolic rates, food consumption and thermoregulation in seasonal acclimatization of the Cape porcupine *Hystrix africaeaustralis*. Oecologia 83:197–200.

Hammel, H. T., T. J. Dawson, R. M. Abrams & H. T. Andersen. 1968. Total calorimetric measurements on *Citellus lateralis* in hibernation. Physiological Zoology 41:341–357.

Hansen, S. & D. M. Lavigne. 1997. Temperature effects on the breeding distribution of grey seals (*Halichoerus grypus*). Physiological Zoology 70:436–443.

Hansen, T. F. 1997. Stabilizing selection and the comparative analysis of adaptation. Evolution 51:1341–1351.

Hansen, T. F. & S. H. Orzack. 2005. Assessing current adaptation and phylogenetic inertia as explanations of trait evolution: The need for controlled comparisons. Evolution 59:2063–2072.

Hansen, T. F., J. Pienaar & S. H. Orzack. 2008. A comparative method for studying adaptation to a randomly evolving environment. Evolution 62:1965–1977.

Hanski, I. & A. Kaikusalo. 1989. Distribution and habitat selection of shrews in Finland. Annales Zoologici Fennici 26:339–348.

Harlow, H. J. 1981a. Metabolic adaptations to prolonged food deprivation by the American badger *Taxidea taxus*. Physiological Zoology 54:276–284.

———. 1981b. Torpor and other physiological adaptations of the badger (*Taxidea taxus*) to cold environments. Physiological Zoology 54:267–275.

Harshman, J., E. L. Braun, M. J. Braun, C. J. Huddleston, R. C. K. Bowie, J. L. Chojnowski, S. J. Hackett, K.-L. Han, R. T. Kimball, B. D. Marks, K. J. Miglia, W. S. Moore, S. Reddy, F. H. Sheldon, D. W. Steadman, S. J. Steppan, C. C. Witt & T. Yuri. 2008. Phylogenomic evidence for multiple losses of flight in ratite birds. Proceedings of the National Academy of Sciences, USA 105:13462–13467.

Hart, J. S. 1956. Seasonal changes in insulation of the fur. Canadian Journal of Zoology 34:53–57.

———. 1971. Rodents. In Comparative Physiology of Thermoregulation, vol. 2, edited by G. C. Whittow, 1–149. Academic Press, New York.

Hart, J. S. & O. Heroux. 1953. A comparison of some seasonal and temperature-induced changes in *Peromyscus*: Cold resistance, metabolism, and pelage insulation. Canadian Journal of Zoology 31:528–534.

Hart, J. S. & L. Irving. 1959. The energetics of harbor seals in air and water with special consideration of seasonal changes. Canadian Journal of Zoology 37:447–457.

Hart, J. S., H. Pohl & J. S. Tener. 1965. Seasonal acclimatization in varying hare (*Lepus americanus*). Canadian Journal of Zoology 43:731–744.

Hartman, F. A. 1961. Locomotor mechanisms of birds. Smithsonian Miscellaneous Collections 143:1–91.

———. 1963. Some flight mechanisms of bats. Ohio Journal of Science 63:59–65.

Harvey, P. H., M. D. Pagel & J. A. Rees. 1991. Mammalian metabolism and life histories. American Naturalist 137:556–566.

Hayes, J. P. 2010. Metabolic rates, genetic constraints, and the evolution of endothermy. Journal of Evolutionary Biology 23:1868–1877.

Hayes, J. P. & T. Garland. 1995. The evolution of endothermy—testing the aerobic capacity model. Evolution 49:836–847.

Hayes, J. P., T. Garland & M. R. Dohm. 1992. Individual variation in metabolism and reproduction of *Mus*: Are energetics and life history linked? Functional Ecology 6:5–14.

Hayssen, V. 1984. Basal metabolic rate and the intrinsic rate of increase: An empirical and theoretical reexamination. Oecologia 64:419–421.

Heath, J. E. 1968. The origins of thermoregulation. In Evolution and the Environment, edited by E. T. Drake, 259–278. Yale University Press, New Haven.

Heath, M. E. & H. T. Hammel. 1986. Body temperature and rate of O_2 consumption in Chinese pangolins. American Journal of Physiology 250:R377–R382.

Heather, B. & H. Robertson. 1966. Field Guide to the Birds of New Zealand. Viking (Penguin Books), Auckland, New Zealand.

Heinrich, B. 1981. Ecological and evolutionary perspectives. In Insect Thermoregulation, edited by B. Heinrich, 236–302. Wiley, New York.

Hennemann, W. W. 1983. Relation among body mass, metabolic rate, and the intrinsic rate of natural increase in mammals. Oecologia 56:104–108.

Herreid, C. F. 1963. Temperature regulation and metabolism in Mexican freetail bats. Science 142:1573–1574.

Herreid, C. F. II & B. Kessel. 1967. Thermal conductance in birds and mammals. Comparative Biochemistry and Physiology 21:405–414.

Heusner, A. A. 1991. Size and power in mammals. Journal of Experimental Biology 160:25–54.

Hibbard, C. W., D. E. Ray, D. E. Savage, D. W. Taylor & J. E. Guilday. 1965. Quaternary mammals of North America. In The Quaternary of the United States, edited by H. E. Wright & D. G. Frey, 509–525. Princeton University Press, Princeton, New Jersey.

Hickling, R., D. B. Roy, J. K. Hill, R. Fox & C. D. Thomas. 2006. The distributions of a wide range of taxonomic groups are expanding polewards. Global Change Biology 12:450–455.

Hiebert, S. M. 1990. Energy costs and temporal organization of torpor in the rufous hummingbird (*Selasphorus rufus*). Physiological Zoology 63:1082–1097.

Hilderbrand, G. V., C. C. Schwartz, C. T. Robbins, M. E. Jacoby, T. A. Hanley, S. M. Arthur & C. Servheen. 1999. The importance of meat, particularly salmon, to body size, population productivity, and conservation of North American brown bears. Canadian Journal of Zoology 77:132–138.

Hildwein, G. 1970. Capacités thermorégulatrices d'un mammifère insectore primitif, le tenrec; leurs variations saisonniès. Archives des Sciences Physiologiques 24:55–71.

———. 1972. Métabolisme énergétique de quelques mammifères et oiseaux de la forêt équatoriale. Archives des Sciences Physiologiques 26:387–400.

———. 1974. Resting metabolic rate in pangolins and squirrels of equatorial rain forest. Archives des Sciences Physiologiques 28:183–195.

Hill, R. W. 1975. Daily torpor in *Peromyscus leucopus* on an adequate diet. Comparative Biochemistry and Physiology A 51:413–423.

Hillenius, W. J. 1994. Turbinates in therapsids: Evidence for late Permian origins of mammalian endothermy. Evolution 48:207–229.

Hillenius, W. J. & J. A. Ruben. 2004. The evolution of endothermy in terrestrial vertebrates: Who? when? why? Physiological and Biochemical Zoology 77:1019–1042.

Hinds, D. S. & R. E. MacMillen. 1985. Scaling of energy metabolism and evaporative water loss in heteromyid rodents. Physiological Zoology 58:282–298.

Hock, R. J. 1960. Seasonal variations in physiologic functions of Arctic ground squirrels and black bears. Bulletin of the Museum of Comparative Zoology 124:155–171.

Hoffmann, R. & R. Prinzinger. 1984. Torpor und Nahrungsausnutzung bei 4 Mausvogelarten (Coliiformes). Journal für Ornithologie 125:225–237.

Hofman, M. A. 1983. Energy metabolism, brain size, and longevity in mammals. Quarterly Review of Biology 58:495–512.

Hooper, E. T. & M. E. Hilali. 1972. Temperature regulation and habits in two species of jerboa, genus *Jaculus*. Journal of Mammalogy 53:574–593.

Hopson, J. A. 1973. Endothermy, small size, and the origin of mammalian reproduction. American Naturalist 107:446–452.

Hudson, J. W. 1965. Temperature regulation and torpidity in the pygmy mouse, *Baiomys taylori*. Physiological Zoology 38:243–254.

Hudson, J. W. & S. L. Kinzey. 1966. Temperature regulation and metabolic rhythms in populations of the house sparrow, *Passer domesticus*. Comparative Biochemistry and Physiology 17:203–217.

Hudson, J. W. & J. A. Rummel. 1966. Water metabolism and temperature regulation of primitive heteromyids *Liomys salvani* and *Liomys irroratus*. Ecology 47:345–354.

Hummel, J., C. T. Gee, K.-H. Südekum, P. M. Sander, G. Nogge & M. Clauss. 2008. In vitro digestibility of fern and gymnosperm foliage: Implications for sauropod feeding ecology and diet selection. Proceedings of the Royal Society of London B 275:1015–1021.

Humphrey, S. R. 1974. Zoogeography of the nine-banded armadillo (*Dasypus novemcinctus*) in the United States. BioScience 24:457–462.

Hussell, D. J. T. 1972. Factors affecting clutch size in Arctic passerines. Ecological Monographs 42:317–364.

Huston, M. A. & S. Wolverton. 2009. The global distribution of net primary productivity: Resolving the paradox. Ecological Monographs 79:343–377.

Hyvärinen, H. 1984. Winter strategy of voles and shrews in Finland. In Winter Ecology of Small Mammals, edited by J. F. Merritt, 139–148. Special Publications of the Carnegie Museum of Natural History, 10. Carnegie Museum of Natural History, Pittsburgh.

Ibáñez, C., J. Juste, J. L. García-Mudarra & P. T. Agirre-Mendi. 2001. Bat predation on nocturnally migrating birds. Proceedings of the National Academy of Sciences, USA 98:9700–9702.

Innes, S. & D. M. Lavigne. 1991. Do cetaceans really have elevated metabolic rates? Physiological Zoology 64:1130–1134.

IPCC. 2007. Summary for policymakers. In Climate Change: The Physical Science Basis. Contribution of working group 1 to the fourth assessment report of the Intergovernmental Panel on Climate Change, edited by S. Solomon et al. Cambridge University Press, Cambridge.

Irestedt, M. & J. I. Olson. 2008. The division of the major songbird radiation into Passerida and "core Corvoidea" (Aves: Passeriformes)—the species trees vs. gene trees. Zoologica Scripta 37:305–313.

Iriarte, J. A., W. L. Franklin, W. E. Johnston & K. H. Redford. 1990. Biogeographic variation of food habits and body size of the American puma. Oecologia 85:185–190.

Irvine, A. B. 1983. Manatee metabolism and its influence on distribution in Florida. Biological Conservation 25:315–334.

Irving, L. & J. S. Hart. 1957. The metabolism and insulation of seals as bare-skinned mammals in cold water. Canadian Journal of Zoology 35:497–511.

Irving, L., O. M. Solandt, D. Y. Solandt & K. C. Fisher. 1935. The respiratory metabolism of the seal and its adjustment to living. Journal of Cellular and Comparative Physiology 7:137–151.

Iversen, J. A. & J. Krog. 1973. Heat production and body surface area in seals and sea otters. Norwegian Journal of Zoology 21:51–54.

Iverson, S. L. & B. N. Turner. 1974. Winter weight dynamics in *Microtus pennsylvanicus*. Ecology 55:1030–1041.

Jaeger, E. C. 1948. Does the Poor-will "hibernate"? Condor 50:45–50.

———. 1949. Further observations on the hibernation of the Poor-will. Condor 51:105–109.

James, F. C. 1970. Geographic size variation in birds and its relationship to climate. Ecology 51:365–390.

Janzen, D. H. 1978. Complications in interpreting the chemical defenses of trees against tropical arboreal plant-eating vertebrates. In The Ecology of Arboreal Folivores, edited by G. G. Montgomery, 73–84. National Zoological Park, Washington, DC.

Jarvis, J. U. M. 1981. Eusociality in a mammal—cooperative breeding in naked mole-rat *Heterocephalus glaber* colonies. Science 212:571–573.

Jarvis, J. U. M., M. J. O'Riain, N. C. Bennett & P. W. Sherman. 1994. Mammalian eusociality: A family affair. Trends in Ecology and Evolution 9:47–51.

Jessop, T. S., T. Madsen, J. Sumner, H. Rudiharto, J. A. Phillips & C. Ciofi. 2006. Maximum body size among insular Komodo dragon populations covaries with large prey density. Oikos 112:422–429.

Jetz, W., R. P. Freckleton & A. E. McKechnie. 2008. Environment, migratory tendency, phylogeny and basal metabolic rate in birds. PLoS One 3:1–9.

Johansson, B. 1957. Some biochemical and electrocardiographic data on the badger. Acta Zoologica (Stockholm) 38:205–218.

Jones, D. L. & L. C.-H. Wang. 1976. Metabolic and cardiovascular adaptations in the western chipmunks, genus *Eutamias*. Journal of Comparative Physiology 105:219–231.

Kalcounis-Rüppell, M. C. 2007. Relationship of basal metabolism and life history attributes in Neotomine-Peromyscine rodents (Cricetidae: Neotominae). Ecoscience 14:347–356.

Kandu, L. L. 2005. Winter energetics of Virginia opossums *Didelphis virginiana* and implications for the species' northern distributional limit. Ecography 28:731–744.

Karasov, W. H. & C. Martínez del Rio. 2007. Physiological Ecology: How Animals Process Energy, Nutrients, and Toxins. Princeton University Press, Princeton, New Jersey.

Kasting, N. W., S. A. L. Adderley, T. Safford & K. G. Hewlett. 1989. Thermoregulation in beluga (*Delphinapterus leucas*) and killer (*Orcinus orca*) whales. Physiological Zoology 62:687–701.

Kemp, T. S. 2005. The origin and evolution of mammals. Oxford University Press, Oxford.

———. 2006. The origin of mammalian endothermy: A paradigm for the evolution of complex biological structure. Zoological Journal of the Linnean Society 147:473–488.

Kendeigh, S. C., V. R. Dol'nik & V. M. Gavrilov. 1977. Avian energetics. In Granivorous Birds in Ecosystems, edited by J. Pinowski & S. C. Kendeigh, 127–204. Cambridge University Press, New York.

Kersten, M., L. W. Bruinzeel, P. Wiersma & T. Piersma. 1998. Reduced metabolic rate of migratory waders wintering in coastal Africa. Ardea 86:71–80.

Kingdon, J. 1974. East African Mammals, vol. IIB. Academic Press, New York.

———. 1977. East African Mammals, vol. IIIA. Academic Press, New York.

Kirkwood, J. K. 1983. A limit to metabolisable energy intake in mammals and birds. Comparative Biochemistry and Physiology A 75:1–3.

Klaassen, M. 1995. Moult and basal metabolic costs in males of two subspecies of stonechats: The European *Saxicola torquata rubicula* and the East African *S. t. axillaris*. Oecologia 104:424–432.

Klaassen, M., M. Kersten & F. J. Ens. 1990. Energetic requirements for maintenance and premigratory body mass gain of waders wintering in Africa. Ardea 78:209–220.

Kleiber, M. 1932. Body size and metabolism. Hilgardia 6:315–353.

———. 1961. Fire of Life: An Introduction to Animal Energetics. Wiley, New York.

Klein, H. 1972. Untersuchungen zur Ökologie und zur verhaltens- und stoffwechselphysiologischen Anpassung von *Talpa europaea* (Linné 1758) an das Mikroklima seines Baues. Zeitschrift für Säugetierkunde 37:16–37.

Köhler, M. 2010. Fast or slow? The evolution of life history traits associated with insular dwarfing. In Islands and Evolution, edited by V. Pérez-Mellado & C. Ramon, 261–280. Recerca 19. Institut Menorqui d'Estudis, Menorca, Spain.

Köhler, M. & S. Moyà-Solà. 2009. Slow life history and physiological plasticity: Survival strategies of a large mammal in a resource-poor environment. Proceedings of the National Academy of Sciences, USA 106:20354–20358.

Kolb, H. H. 1978. Variation in the size of foxes in Scotland. Biological Journal of the Linnean Society 10:291–304.

Kolokotrones, T., V. Savage, E. J. Deeds & W. Fontana. 2010. Curvature in metabolic scaling. Nature 464:753–756.

Kooyman, G. L., Y. Cherel, Y. Le Maho, J. P. Croxall, P. H. Thorson, V. Ridoux & C. A. Kooyman. 1992. Diving behaviour and energetics during foraging cycles in King Penguins. Ecological Monographs 62:143–163.

Kooyman, G. L., D. H. Kerem, W. B. Campbell & J. J. Wright. 1973. Pulmonary gas exchange in freely diving Weddell seals, *Leptonychotes weddelli*. Respiration Physiology 17:283–290.

Korhonen, K. 1980. Microclimate in the snow burrows of willow grouse (*Lagopus lagopus*). Annales Zoologici Fennici 17:5–9.

Körtner, G., R. M. Brigham & F. Geiser. 2001. Torpor in free-ranging tawny frogmouths (*Podargus strigoides*). Physiological and Biochemical Zoology 74:789–797.

Koteja, P. 1991. On the relation between basal and field metabolic rates in birds and mammals. Functional Ecology 5:56–64.

Krebs, C. J. 1972. Ecology: The experimental analysis of distribution and abundance. Harper & Row, New York.

Król, E. 1994. Metabolism and thermoregulation in the eastern hedgehog *Erinaceus concolor*. Journal of Comparative Physiology B 164:503–507.

Krzanowski, A. 1961. Weight dynamics of bats wintering in a cave at Pulawy (Poland). Acta Theriologica 4:242–264.

Kurta, A. & M. Ferkin. 1991. The correlation between demography and metabolic rate: A test using the beach vole (*Microtus breweri*) and the meadow vole (*Microtus pennsylvanicus*). Oecologia 87:102–105.

Labra, A., J. Pienaar & T. F. Hansen. 2009. Evolution of thermal physiology in *Liolaemus* lizards: Adaptation, phylogenetic inertia, and niche tracking. American Naturalist 174:204–220.

Lack, D. 1947. The significance of clutch size. Parts I and II. Ibis 89:302–352.

———. 1948. The significance of clutch size. Part III. Ibis 90:25–45.

Lane, J. E., R. M. Brigham & D. L. Swanson. 2004. Daily torpor in free-ranging Whip-poor-wills (*Caprimulgus vociferous*). Physiological and Biochemical Zoology 77:297–304.

Langvatn, R. & S. D. Albon. 1986. Geographic clines in body weight of Norwegian red deer: A novel explanation of Bergmann's rule? Holarctic Ecology 9:285–293.

Lasiewski, R. C. & W. R. Dawson. 1964. Physiological responses to temperature in the common nighthawk. Condor 66:477–490.

———. 1967. A re-examination of the relation between standard metabolic rate and body weight in birds. Condor 69:13–23.

Lasiewski, R. C., W. W. Weathers & M. H. Bernstein. 1967. Physiological responses of the giant hummingbird, *Patagona gigas*. Comparative Biochemistry and Physiology 23:797–813.

Lavigne, D. M., W. Barchard, S. Innes & N. A. Øritsland. 1982. Pinniped energetics. FAO Fisheries Series 5:191–235.

Lavigne, D. M., S. Innes, G. A. J. Worthy, K. M. Kovacs, O. J. Schmitz & J. P. Hickie. 1986. Metabolic rates of seals and whales. Canadian Journal of Zoology 64:279–284.

Layne, J. N. & J. R. Redmond. 1959. Body temperatures of the least shrew, *Cryptotis parva floridiana* (Merriam, 1895). Säugetierkundliche Mitteilungen 7:169–172.

Lehman, S. M., M. Mayor & P. C. Wright. 2005. Ecogeographic size variations in sifakas: A test of the resource seasonality and resource quality hypotheses. American Journal of Physical Anthropology 126:318–328.

Leitner, P. 1966. Body temperature, oxygen consumption, heart rate and shivering in the California mastiff bat, *Eumops perotis*. Comparative Biochemistry and Physiology 19:431–443.

Lenoir, J., J. C. Gégout, P. A. Marquet, P. de Ruffray & H. Brisse. 2008. A significant upward shift in plant species optimum elevation during the 20th century. Science 320:1768–1771.

Lepage, D. & P. Lloyd. 2004. Avian clutch size in relation to rainfall seasonality and stochasticity along an aridity gradient across South Africa. Ostrich 75:259–268.

Licht, P. & P. Leitner. 1967. Physiological responses to high environmental temperatures in three species of microchiropteran bats. Comparative Biochemistry and Physiology 22:371–387.

Lillegraven, J. A. 1976. Biological considerations of the marsupial-placental dichotomy. Evolution 29:707–722.

———. 1985. Use of the term trophoblast for tissues in therian mammals. Journal of Morphology 183:293–299.

Lillegraven, J. A., S. D. Thompson, B. K. McNab & J. L. Patton. 1987. The origin of eutherian mammals. Biological Journal of the Linnean Society 32:281–336.

Lindstedt, S. L. 1980. Regulated hypothermia in the desert shrew. Journal of Comparative Physiology 137:173–176.

Lindstedt, S. L. & M. S. Boyce. 1985. Seasonality, fasting endurance, and body size in mammals. American Naturalist 125:873–878.

Lindström, Å. & A. Kvist. 1995. Maximum energy intake is proportional to basal metabolic rate in passerine birds. Proceedings of the Royal Society of London B 261:337–343.

Lindström, A., A. Kvist, T. Piersma, A. Dekinga & M. W. Dietz. 2000. Avian pectoral muscle size rapidly tracks body mass changes during flight, fasting and fueling. Journal of Experimental Biology 203:913–919.

Lessa, E. P. 1990. Morphological evolution of subterranean mammals: Integrating structural, functional, and ecological perspectives. In Evolution of Subterranean Mammals at the Organismal and Molecular Levels, edited by E. Nevo & O. A. Reig, 211–230. Wiley-Liss, New York.

Lister, A. M. 1989. Rapid dwarfing of red deer on Jersey in the Last Interglacial. Nature 342:539–542.

———. 1993. Mammoths in miniature. Nature 362:288–289.

Livezey, B. C. 1992. Morphological corollaries and ecological implications of flightlessness in the Kakapo (Psittaciformes: *Strigops habroptilus*). Journal of Morphology 213:105–145.

Lloyd, P. 1999. Rainfall as a breeding stimulus and clutch size determinant in South African arid-zone birds. Ibis 141:637–643.

Lomolino, M. V. 1985. Body size of mammals on islands: The island rule reexamined. American Naturalist 125:310–316.

―――. 2005. Body size evolution in insular vertebrates: Generality of the island rule. Journal of Biogeography 32:1683–1699.

Lovegrove, B. G. 1986a. The metabolism of social subterranean rodents: Adaptation to aridity. Oecologia 69:551–555.

―――. 1986b. Thermoregulation of the subterranean rodent genus *Bathyergus* (Bathyergidae). South African Journal of Zoology 21:283–288.

―――. 1987. Thermoregulation in the subterranean rodent *Georychus capensis* (Rodentia: Bathyergidae). Physiological Zoology 60:174–180.

―――. 2000. The zoogeography of mammalian basal metabolic rate. American Naturalist 156:201–219.

―――. 2001. The evolution of body armor in mammals: Plantigrade constraints of large body size. Evolution 55:1464–1473.

―――. 2009. Age at first reproduction and growth rate are independent of basal metabolic rate in mammals. Journal of Comparative Physiology B 179:391–401.

Lynch, G. R., F. D. Vogt & H. R. Smith. 1978. Seasonal study of spontaneous daily torpor in the white-footed mouse, *Peromyscus leucopus*. Physiological Zoology 51:289–299.

Macari, M., D. L. Ingram & M. J. Dauncey. 1983. Influence of thermal and nutritional acclimation on body temperatures and metabolic rate. Comparative Biochemistry and Physiology A 74:549–553.

MacMahon, T. 1973. Size and shape in biology. Science 179:1201–1204.

MacMillen, R. E. 1965. Aestivation in the cactus mouse, *Peromyscus eremicus*. Comparative Biochemistry and Physiology 16:227–248.

MacMillen, R. E., R. V. Baudinette & A. K. Lee. 1972. Water economy and energy metabolism of sandy inland mouse, *Leggadina hermannsburgensis*. Journal of Mammalogy 53:529–539.

MacMillen, R. E. & A. K. Lee. 1970. Energy metabolism and pulmocutaneous water loss of Australian hopping mice. Comparative Biochemistry and Physiology 35:355–369.

Maehr, D. S., R. C. Belden, E. D. Land & L. Wilkins. 1990. Food habits of panthers in southwest Florida. Journal of Wildlife Management 54:420–423.

Maloiy, G. M. O., J. M. Z. Kamu, A. Shkolnik, M. Meir & R. Arieli. 1982. Thermoregulation and metabolism in a small desert carnivore: The fennec fox (*Fennicus zerda*) (Mammalia). Journal of Zoology, London 198:279–291.

Marshall, J. T., Jr. 1955. Hibernation in captive goatsuckers. Condor 57:129–134.

Marshall, L. G., R. F. Butler, R. E. Drake, G. H. Curtis & R. H. Telford. 1979. Calibration of the Great American Interchange. Science 204:272–279.

Marshall, L. G., R. Pasqual, G. H. Curtis & R. E. Drake. 1977. South American geochronology: Radiometric time scale for Middle to Late Tertiary mammal-bearing horizons in Patagonia. Science 195:1325–1328.

Marshall, L. G., S. D. Webb, J. J. Sepkoski & D. M. Raup. 1982. Mammalian evolution and the Great American Interchange. Science 215:1351–1357.

Martin, R. D. 1981. Relative brain size and basal metabolic rate in terrestrial vertebrates. Nature 293:57–60.

―――. 1998. Comparative aspects of human brain evolution: Scaling, energy costs and confounding variables. In The Origin and Diversification of Language, edited by N. G. Jablonski & L. C. Aiello, 35–68. Memoirs of the California Academy of Sciences. California Academy of Sciences, San Francisco.

Martin, T. E. 1996. Life history evolution in tropical and south temperate birds: What do we really know? Journal of Avian Biology 27:263–272.

Martin, T. E., P. R. Martin, C. R. Olson, B. J. Heidinger & J. J. Fontaine. 2000. Parental care and clutch sizes in North and South American birds. Science 287:1482–1485.

Martuscelli, P. 1995. Avian predation by the round-eared bat (*Tonatia bidens*, Phyllostomidae) in the Brazilian Atlantic forest. Journal of Tropical Ecology 11:461–464.

Masman, D. & M. Klaassen. 1987. Energy expenditure during free flight in trained and free-living Eurasian kestrels (*Falco tinnunculus*). Auk 104:603–616.

Mathias, M. L., A. C. Nunes, C. C. Marques, I. Sousa, M. G. Ramalhindo, J. C. Auffray, J. Catalan & J. Brittan-Davidian. 2004. Adaptive energetics in house mice, *Mus musculus*, from the island of Porto Santo (Madeira Archipelago, North Atlantic). Comparative Biochemistry and Physiology A 137:703–709.

May, M. L. 1979. Insect thermoregulation. Annual Review of Entomology 24:313–349.

Mayr, E. 1956. Geographical character gradients and climatic adaptation. Evolution 10:105–108.

Mayr, E. & J. Diamond. 2001. The Birds of Northern Melanesia: Speciation, Ecology, and Biogeography. Oxford University Press, New York.

McCabe, T. T. & B. D. Blanchard. 1950. Three Species of *Peromyscus*. Rood Associates, Santa Barbara, California.

McFarland, W. N. & W. A. Wimsatt. 1969. Renal function and its relation to the ecology of the vampire bat, *Desmodus rotundus*. Comparative Biochemistry and Physiology 28:985–1006.

McKechnie, A E., R. A. M. Ashdown, M. B. Christian & R. M. Brigham. 2007. Torpor in an African caprimulgid, the freckled nightjar *Caprimulgus tristigma*. Journal of Avian Biology 38:261–266.

McKechnie, A. E., G. Kortner & B. G. Lovegrove. 2004. Rest-phase thermoregulation in free-ranging White-backed Mousebirds. Condor 106:143–149.

McKechnie, A. E. & B. G. Lovegrove. 2001. Heterothermic responses in the speckled mouse-bird (*Colius striatus*). Journal of Comparative Physiology B 171:507–518.

McKechnie, A. E. & B. O. Wolf. 2004. The allometry of avian basal metabolic rate: Good predictions need good data. Physiological and Biochemical Zoology 77:502–521.

McKinnon, L., P. A. Smith, E. Nol, J. L. Martin, F. I. Doyle, K. F. Abraham, H. G. Gilchrist, R. I. G. Morrison & J. Bêty. 2010. Lower predation risk for migratory birds at high latitudes. Science 327:326–327.

McNab, B. K. 1966. The metabolism of fossorial rodents: A study of convergence. Ecology 47:712–733.

———. 1969. The economics of temperature regulation in Neotropical bats. Comparative Biochemistry and Physiology 31:227–268.

———. 1971. On the ecological significance of Bergmann's rule. Ecology 52:845–854.

———. 1973. Energetics and the distribution of vampires. Journal of Mammalogy 54:131–144.

———. 1974. The behavior of temperate cave bats in a subtropical environment. Ecology 55:943–958.

———. 1978a. The evolution of endothermy in the phylogeny of mammals. American Naturalist 112:1–21.

———. 1978b. The comparative energetics of Neotropical marsupials. Journal of Comparative Physiology 125:115–128.

————. 1978c. Energetics of arboreal folivores: Physiological problems and ecological consequences of feeding on an ubiquitous food supply. In The Ecology of Arboreal Folivores, edited by G. G. Montgomery, 153–162. Smithsonian Institution Press, Washington, DC.

————. 1979a. Climatic adaptation in the energetics of heteromyid rodents. Comparative Biochemistry and Physiology A 62:813–820.

————. 1979b. The influence of body size on the energetics and distribution of fossorial and burrowing mammals. Ecology 60:1010–1021.

————. 1980a. On estimating thermal conductance in endotherms. Physiological Zoology 53:145–156.

————. 1980b. Food habits, energetics, and the population biology of mammals. American Naturalist 116:106–124.

————. 1980c. Energetics and the limit to a temperate distribution in armadillos. Journal of Mammalogy 61:606–627.

————. 1982. The physiological ecology of South American mammals. In Mammalian Biology in South America, edited by M. A. Mares & H. H. Genoways, 187–207. Special Publications 6. Pymanuning Laboratory of Ecology, University of Pittsburgh, Pittsburgh, Pennsylvania.

————. 1983. Energetics, body size, and the limits to endothermy. Journal of Zoology, London 199:1–29.

————. 1984. Physiological convergence amongst ant-eating and termite-eating mammals. Journal of Zoology, London 203:485–510.

————. 1986a. The influence of food habits on the energetics of eutherian mammals. Ecological Monographs 56:1–19.

————. 1986b. Food habits, energetics, and the reproduction of marsupials. Journal of Zoology, London 208:595–614.

————. 1988a. Energy conservation in a tree-kangaroo (*Dendrolagus matschiei*) and the red panda (*Ailurus fulgens*). Physiological Zoology 61:280–292.

————. 1988b. Food habits and the basal rate of metabolism in birds. Oecologia 77:343–349.

————. 1989. On the selective persistence of mammals in South America. In Advances in Neotropical Mammalogy, edited by K. H. Redford & J. F. Eisenberg, 605–614. Sandhill Crane Press, Gainesville, Florida.

————. 1991. The energy expenditure of shrews. In The Biology of the Soricidae, edited by J. S. Findley & T. L. Yates, 35–45. Special Publications of the Museum of Southwestern Biology, 1. Museum of Southwestern Biology, University of New Mexico, Albuquerque, New Mexico.

————. 1992a. Rate of metabolism in the termite-eating sloth bear (*Ursus ursinus*). Journal of Mammalogy 73:168–173.

————. 1992b. The comparative energetics of rigid endothermy: The Arvicolidae. Journal of Zoology, London 227:586–606.

————. 1994a. Energy conservation and the evolution of flightlessness in birds. American Naturalist 144:628–642.

————. 1994b. Resource use and the occurrence of land and freshwater vertebrates on oceanic islands. American Naturalist 144:643–660.

————. 1995. Energy expenditure and conservation in frugivorous and mixed-diet carnivorans. Journal of Mammalogy 76:206–222.

————. 1996. Metabolism and temperature regulation of kiwis (Apterygidae). Auk 113:687–692.

————. 1997. On the utility of uniformity in the definition of basal rate of metabolism. Physiological Zoology 70:718–720.

————. 1999. On the comparative and evolutionary significance of total and mass-specific rates of metabolism. Physiological and Biochemical Zoology 72:642–644.

————. 2000a. Energy constraints on carnivore diet. Nature 907:584.

————. 2000b. The influence of body mass, climate, and distribution on the energetics of South Pacific pigeons. Comparative Biochemistry and Physiology A 127:309–329.

————. 2000c. The standard energetics of mammalian carnivores: Felidae and Hyaenidae. Canadian Journal of Zoology 78:1–13.

————. 2001. The energetics of toucans, a barbet, and a hornbill: Implications for avian frugivory. Auk 118:916–933.

————. 2002a. The Physiological Ecology of Vertebrates: A View from Energetics. Cornell University Press, Ithaca, New York.

————. 2002b. Minimizing energy expenditure facilitates vertebrate persistence on oceanic islands. Ecology Letters 5:693–704.

————. 2002c. Short-term energy conservation in endotherms in relation to body mass, habits, and environment. Journal of Thermal Biology 27:459–466.

————. 2003a. The energetics of New Zealand's ducks. Comparative Biochemistry and Physiology A 135:229–247.

————. 2003b. Standard energetics of phyllostomid bats: The inadequacies of phylogenetic-contrast analyses. Comparative Biochemistry and Physiology A 135:357–368.

————. 2003c. Ecology shapes bird bioenergetics. Nature 426:620–621.

————. 2005a. Food habits and the evolution of energetics in birds of paradise (Paradisaeidae). Journal of Comparative Physiology B 175:117–132.

————. 2005b. Uniformity in the BMR of marsupials: Its causes and consequences. Revista Chilena de Historia Natural 78:183–198.

————. 2005c. Ecological factors influence energetics in the Order Carnivora. Acta Zoologica Sinica 51:535–545.

————. 2006a. The evolution of energetics in eutherian "insectivorans:" An alternate approach. Acta Theriologica 51:113–128.

————. 2006b. The energetics of reproduction in endotherms and its implication for their conservation. Integrative and Comparative Biology 46:1159–1168.

————. 2008a. An analysis of the factors that influence the level and scaling of mammalian BMR. Comparative Biochemistry and Physiology A 151:5–28.

————. 2008b. The comparative energetics of New Guinean cuscuses (Metatheria: Phalangeridae). Journal of Mammalogy 89:1145–1151.

————. 2008c. Physiological ecology. In Encyclopedia of Ecology, vol. 4., edited by S. E. Jørgensen & B. D. Fath, 2744–2751. Elsevier, Oxford.

————. 2009a. Ecological factors affect the level and scaling of avian BMR. Comparative Biochemistry and Physiology A 152:22–45.

————. 2009b. Energy expenditure cannot be effectively analyzed with phylogenetically based techniques. In Molecules to Migration: The Pressures of Life, edited by S. Morris & A. Vosloo, 621–626. Medimond Editore, Bologna, Italy.

————. 2009c. Resources and energetics determined dinosaur body size. Proceedings of the National Academy of Sciences, USA 106:12184–12188.

———. 2010. Geographic and temporal correlations of mammalian size reconsidered: A resource rule. Oecologia 164:13–23.

McNab, B. K. & W. Auffenberg. 1976. The effect of large body size on the temperature regulation of the Komodo dragon, *Varanus komodoensis*. Comparative Biochemistry and Physiology A 55:345–350.

McNab, B. K. & F. J. Bonaccorso. 1995. The energetics of pteropodid bats. In Ecology, Evolution, and Behaviour of Bats, edited by P. A. Racey & S. M. Swift, 111–122. Symposia of the Zoological Society of London, 67. Clarendon Press, Oxford.

———. 2001. The metabolism of New Guinean pteropodid bats. Journal of Comparative Physiology B 171:1–13.

McNab, B. K. & J. F. Eisenberg. 1989. Brain size and its relation to the rate of metabolism in mammals. American Naturalist 133:157–167.

McNab, B. K. & H. I. Ellis. 2006. Flightless rails endemic to islands have lower energy expenditures and clutch sizes than flighted rails on islands and continents. Comparative Biochemistry and Physiology A 145:295–311.

McNab, B. K. & M. Köhler. Brain size and the BMR of bats. Unpublished manuscript.

McNab, B. K. & P. R. Morrison. 1963. Body temperature and metabolism of subspecies of *Peromyscus* from arid and mesic environments. Ecological Monographs 33:63–82.

McNab, B. K. & C. A. Salisbury. 1995. Energetics of New Zealand's temperate parrots. New Zealand Journal of Zoology 22:339–349.

McNab, B. K. & P. C. Wright. 1987. Temperature regulation and oxygen consumption in the Philippine tarsier (*Tarsius syrichta*). Physiological Zoology 60:596–600.

McNamara, J. M., Z, Barta, M. Wikelski & A. I. Houston. 2008. A theoretical investigation of the effect of latitude on avian life. American Naturalist 172:331–345.

Mead, J. L., D. W. Steadman, S. H. Bedford, C. J. Bell & M. Spriggs. 2002. New extinct mekosuchine crocodile from Vanuatu, South Pacific. Copeia 2002:632–641.

Medina, A., D. A. Marti & C. J. Bidau. 2007. Subterranean rodents of the genus *Ctenomys* (Caviomorpha, Ctenomyidae) follow the converse to Bergmann's rule. Journal of Biogeography 34:1439–1454.

Meiri, S. & T. Dayan. 2003. On the validity of Bergmann's rule. Journal of Biogeography 30:331–356.

Meiri, S., T. Dayan & D. Simberloff. 2004. Carnivores, biases and Bergmann's rule. Biological Journal of the Linnean Society 81:579–588.

Meiri, S., N. Cooper & A. Purvis. 2008a. The island rule: Made to be broken? Proceedings of the Royal Society of London B 275:141–148.

Meiri, S., E. Meijaard, S. A. Wich, C. P. Groves & K. M. Helgen. 2008b. Mammals of Borneo—small size on a large island. Journal of Biogeography 35:1087–1094.

Merkt, J. R. & C. R. Taylor. 1994. Metabolic switch for desert survival. Proceedings of the National Academy of Sciences, USA 91:12313–12316.

Merola-Zwartjes, M. & J. D. Ligon. 2000. Ecological energetics of the Puerto Rican tody: Heterothermy, torpor, and intra-island variation. Ecology 81:990–1003.

Merriam, C. H. 1894. Laws of temperature control of the geographic distribution of terrestrial animals and plants. National Geographic Magazine 6:229–238.

Merritt, J. F. & D. A. Zegers. 1991. Seasonal thermogenesis and body-mass dynamics of *Clethrionomys gapperi*. Canadian Journal of Zoology 69:2771–2777.

———. 2002. Maximizing survivorship in cold: Thermogenic profiles of non-hibernating mammals. Acta Theriologica 47:221–234.

Mezhzherin, V. A. 1964. Dehnel's phenomenon and its possible explanation. Acta Theriologica 8:95–114.

Mezhzherin, V. A. & G. L. Melnikova. 1966. Adaptive importance of seasonal changes in some morphophysiological indices in shrews. Acta Theriologica 11:503–521.

Millar, J. S. & G. J. Hickling. 1990. Fasting endurance and the evolution of mammalian body size. Functional Ecology 4:5–12.

Miller, K. & L. Irving. 1975. Metabolism and temperature regulation in young harbor seals *Phoca vitulina richardi*. American Journal of Physiology 229:506–511.

Miller, K., M. Rosenmann & P. R. Morrison. 1976. Oxygen uptake and temperature regulation of young harbor seals (*Phoca vitulina richardi*) in water. Comparative Biochemistry and Physiology A 54:105–107.

Montgomery, G. G. & M. E. Sunquist. 1978. Habitat selection and use by two-toed and three-toed sloths. In The Ecology of Arboreal Folivores, edited by G. G. Montgomery, 329–359. National Zoological Park, Washington, DC.

Moreau, R. E. 1944. Clutch size: A comparative study, with reference to African birds. Ibis 86:286–347.

Morrison, P. & B. K. McNab. 1962. Daily torpor in a Brazilian murine opossum (*Marmosa*). Comparative Biochemistry and Physiology 6:57–68.

Morrison, P. R. & F. A. Ryser. 1951. Temperature and metabolism in some Wisconsin mammals. Federation Proceedings 10:93–94.

Morrison, P. R. & W. J. Tietz. 1957. Cooling and thermal conductivity in three small Alaskan mammals. Journal of Mammalogy 38:78–86.

Mueller, P. & J. Diamond. 2001. Metabolic rate and environmental productivity: Well-provisioned animals evolved to run and idle fast. Proceedings of the National Academy of Sciences, USA 98:12550–12554.

Mugaas, J. N. & J. Seidensticker. 1993. Geographic variation of lean body mass and a model of its effect on the capacity of the raccoon to fatten and fast. Bulletin of the Florida Museum of Natural History 36:85–107.

Müller, E. F. 1979. Energy metabolism, thermoregulation and water budget in the slow loris (*Nycticebus coucang*, Boddaert 1785). Comparative Biochemistry and Physiology A 64:109–119.

Müller, E. F. & H. Jaksche. 1980. Thermoregulation, oxygen consumption, heart rate and evaporative water loss in the thick-tailed bushbaby (*Galago crassicaudatus* Geoffroy, 1812). Zeitschrift für Säugetierkunde 45:269–278.

Müller, E. F. & E. Kulzer. 1977. Body temperature and oxygen uptake in the kinkajou (*Potos flavus*, Schreber), a nocturnal tropical carnivore. Archives Internationales de Physiologie et de Biochemie 86:153–163.

Müller, E. F., U. Nieschalk & B. Meier. 1985. Thermoregulation in the slender loris (*Loris tardigradus*). Folia Primatologica 44:216–226.

Murphy, E. C. 1985. Bergmann's rule, seasonality, and geographic variation in body size of house sparrows. Evolution 39:1327–1334.

Nagel, A. & R. Nagel. 1991. How do bats choose optimal temperatures for hibernation? Comparative Biochemistry and Physiology A 99:323–326.

Nagy, K. A. 1987. Field metabolic rate and food requirement scaling in mammals and birds. Ecological Monographs 57:111–128.

———. 1994. Field bioenergetics of mammals: What determines field metabolic rates? Australian Journal of Zoology 42:43–53.

————. 2005. Field metabolic rate and body size. Journal of Experimental Biology 208:1621–1625.

Nagy, K. A., I. A. Girard & T. K. Brown. 1999. Energetics of free-ranging mammals, reptiles, and birds. Annual Review of Nutrition 19:247–277.

Nagy, K. A. & M. H. Knight. 1994. Energy, water, and food use by springbok antelope (*Antidorcas marsupialis*) in the Kalahari Desert. Journal of Mammalogy 75:860–872.

Nespolo, R. F., L. D. Bacigalupe, C. C. Figueroa, P. Koteja & J. C. Opazo. 2011. Using new tools to solve an old problem: The evolution of endothermy in vertebrates. Trends in Ecology and Evolution, 16 May 2011, 1–10. doi: 10.1016/j.tree.2011.04.004.

Nevo, E. 1999. Mosaic evolution of subterranean mammals: Regression, progression, and global convergence. Oxford University Press, Oxford.

Nicoll, M. E. & S. D. Thompson. 1987. Basal metabolic rates and energetics of reproduction in therian mammals: Marsupials and placentals compared. Symposium of the Zoological Society of London 57:7–27.

Noll-Banholzer, U. 1979. Body temperature, oxygen consumption, evaporative water loss and heart rate in the fennec. Comparative Biochemistry and Physiology A 62:585–592.

Nordøy, E. S. & E. S. Blix. 1985. Energy sources in fasting grey seal pups evaluated with computed tomography. American Journal of Physiology 249:R471–R476.

Novakowski, N. S. 1967. Winter bioenergetics of a beaver population in northern latitudes. Canadian Journal of Zoology 45:1107–1110.

Novick, A. & B. A. Dale. 1971. Foraging behavior in fishing bats and their insectivorous relatives. Journal of Mammalogy 52:817–818.

Obst, B. S. & K. A. Nagy. 1992. Field energy expenditures of the southern giant-petrel. Condor 94:801–810.

Ochoa-Acuña, H. G., B. K. McNab & E. H. Miller. 2009. Seasonal energetics of northern phocid seals. Comparative Biochemistry and Physiology A 152:341–350.

Ochocińska, D. & J. R. E. Taylor. 2003. Bergmann's rule in shrews: Geographical variation of body size in Palearctic *Sorex* species. Biological Journal of the Linnean Society 78:365–381.

Okarma, H. & P. Koteja. 1987. Basal metabolic rate in the gray wolf in Poland. Journal of Wildlife Management 51:800–801.

Ostrom, J. H. 1980. The evidence for endothermy in dinosaurs. In A Cold Look at Warm-Blooded Dinosaurs, edited by R. D. Thomas & E. C. Olson, 5–54. Westview Press, Boulder, Colorado.

Padley, D. 1985. Do the life-history parameters of passerines scale to metabolic rate independently of body mass? Oikos 45:285–287.

Padykula, H. A. & J. M. Taylor. 1976. Ultrastructural evidence for loss of trophoblastic layer in chorioallantoic placenta of Australian bandicoots (Marsupialia, Peramelidae). Anatomical Record 186:357–385.

Palacio, C. 1977. Standard metabolism and thermoregulation in three species of lorosoid primates. Master's thesis, University of Florida, Gainesville

Paladino, F. V., M. P. O'Connor & J. R. Spotila. 1990. Metabolism of leatherback turtles, gigantothermy, and thermoregulation of dinosaurs. Nature 344:858–860.

Palmqvist, P. & S. F. Vizcaíno. 2003. Ecological and reproductive constraints of body size in the gigantic *Argentavis magnificens* (Aves, Theratornithidae) from Miocene of Argentina. Ameghiniana 40:379–385.

Pascual, R., M. Archer, E. O. Jaureguizar, J. L. Prado, H. Godthelp & S. J. Hand. 1992. First discovery of monotremes in South America. Nature 377:418–420.

Pascual, R., F. J. Goin, L. Balino & D. E. U. Sauthier. 2002. New data on the Paleocene monotreme *Monotrematum sudamericanum*, and the convergent evolution of triangulate molars. Acta Palaeontologica Polonica 47:487–492.

Patten, M. A. 2007. Geographic variation in calcium and clutch size. Journal of Avian Biology 38:637–643.

Patterson, B. & R. Pascual. 1972. The fossil mammal fauna of South America. In Evolution, Mammals and Southern Continents, edited by A. Keast, F. C. Erk & B. Glass, 247–309. State University of New York Press, Albany.

Paul, G. S. 1988. Physiological, migratorial, climatological, geophysical, survival, and evolutionary implications of Cretaceous polar dinosaurs. Journal of Paleontology 62:640–652.

Peczkis, J. 1994. Implications of body mass estimates for dinosaurs. Journal of Vertebrate Paleontology 14:520–533.

Peiponen, V. A. 1965. On hypothermia and torpidity in the nightjar (*Caprimulgus europaeus* L.). Annales Academiae Scientiarum Fennicae A. IV. Biologica 87:1–15.

———. 1966. The diurnal heterothermy of the nightjar (*Caprimulgus europaeus* L.). Annales Academiae Scientiarum Fennicae A. IV. Biologica 101:1–35.

———. 1970. Body temperature fluctuations in the nightjar (*Caprimulgus e. europaeus* L.) in light conditions of southern Finland. Annales Zoologici Fennici 7:239–250.

Pennycuick, C. J. 1982. The flight of petrels and albatrosses (Procellariiformes), observed in South Georgia and its vicinity. Philosophical Transactions of the Royal Society of London B 300:75–106.

Peterson, C. C., K. A. Nagy & J. Diamond. 1990. Sustained metabolic scope. Proceedings of the National Academy of Sciences, USA 87:2324–2328.

Pettit, T. N., K. A. Nagy, H. I. Ellis & G. C. Whittow. 1988. Incubation energetics of the Laysan albatross. Oecologia 74:546–550.

Phillips, M. J., G. C. Gibb, E. A. Crump & D. Penny. 2010. Tinamous and moa flock together: Mitochondrial genome sequence analysis reveals independent losses of flight among ratites. Systematic Biology 59:90–107.

Piersma, T., L. Bruinzeel, R. Drent, M. Kersten, J. V. der Meer & P. Wiersma. 1996. Variability in basal metabolic rate of a long-distance migrant shorebird (red knot, *Calidris canutus*) reflects shifts in organ sizes. Physiological Zoology 69:191–217.

Piersma, T., N. Cadee & S. Daan. 1995. Seasonality in basal metabolic rate and thermal conductance in a long-distance migrant shorebird, the knot (*Calidris canutus*). Journal of Comparative Physiology B 165:37–45.

Popa-Lisseanu, A. G., A. Delgado-Huertas, M. G. Forero, A. Rodriquez, R. Arlettaz & C. Ibáñez. 2007. Bat's conquest of a formidable foraging niche: The myriads of nocturnally migrating songbirds. PLoS ONE 2:e205. doi:10.1371/journal.pone.0000205.

Porter, W. P. & D. M. Gates. 1969. Thermodynamic equilibria of animals with environment. Ecological Monographs 39:245–270.

Predavec, M. 1997. Variable energy demands in *Pseudomys hermannsburgensis*: Possible ecological consequences. Australian Journal of Zoology 45:85–94.

Prinzinger, R., R. Göpel, A. Lorenz & E. Kulzer. 1981. Body temperature and metabolism in the red-backed mousebird (*Colius castanotus*) during fasting and torpor. Comparative Biochemistry and Physiology A 69:689–692.

Prinzinger, R. & I. Hänssler. 1980. Metabolism-weight relationship in some small nonpasserine birds. Experientia 36:1299–1300.

Prinzinger, R., T. Schäfer & K.-L. Schuchmann. 1992. Energy metabolism, respiratory quotient and breathing parameters in two convergent small bird species: The fork-tailed sunbird *Aethopyga christinae* (Nectariniidae) and the Chilean hummingbird *Sephanoides sephanoides* (Trochilidae). Journal of Thermal Biology 17:71–79.

Pucek, Z. 1963. Seasonal changes in the braincase of some representatives of the genus *Sorex* from the Palearctic. Journal of Mammalogy 44:523–536.

———. 1964. Morphological changes in shrews kept in captivity. Acta Theriologica 8:137–166.

Pujos, F. & R. Salas. 2004. A systematic reassessment and paleogeographic review of fossil Xenarthra from Peru. Bulletin de l'Institut Français d'Études Andines 33:331–377.

Rabosky, D. L. 2009. Ecological limits on clade diversification in higher taxa. American Naturalist 173:622–674.

Raia, P. & S. Meiri. 2006. The island rule in large mammals: Paleontology meets ecology. Evolution 60:1731–1742.

Ralls, K. & P. H. Harvey. 1985. Geographic variation in size and sexual dimorphism of North American weasels. Biological Journal of the Linnean Society 25:119–167.

Rasmussen, D. T. & M. K. Izard. 1988. Scaling of growth and life history traits relative to body size, brain size, and metabolic rate in lorises and galagos (Lorisidae, Primates). American Journal of Physical Anthropology 75:357–367.

Rathburn, G. B. 1979. The social structure and ecology of elephant-shrews. Advances in Ethology 20:1–77.

Redford, K. H. & J. F. Eisenberg. 1992. Mammals of the Neotropics, The Southern Cone. Chile, Argentina, Paraguay. University of Chicago Press, Chicago.

Rensch, B. 1936. Studien über klimatische parallelitat der merkmalsaus bei vogeln und saugern. Archive für Naturgeschicte 5:317–363.

Repasky, R. R. 1991. Temperature and the northern distributions of wintering birds. Ecology 72:2274–2285.

Reynolds, P. S. & R. M. Lee. 1996. Phylogenetic analysis of avian energetics: Passerines and nonpasserines do not differ. American Naturalist 147:735–759.

Rezende, E L., D. L. Swanson, F. F. Novoa & F. Bozinovic. 2002. Passerines versus nonpasserines: So far, no statistical differences in the scaling of avian energetics. Journal of Experimental Biology 205:101–107.

Richard, A. F. & M. E. Nicol. 1987. Female social dominance and basal metabolism in a Malagasy primate, *Propithecus verreauxi*. American Journal of Primatology 12:309–314.

Richardson, P. R. K. 1987. Aardwolf: The most specialized myrmecophagous mammal? South African Journal of Science 83:643–646.

Ricklefs, R. E. 1968. Patterns of growth in birds. Ibis 110:419–451.

———. 1976. Growth rates of birds in humid New World tropics. Ibis 118:179–207.

———. 1980. Geographical variation in clutch size among passerine birds: Ashmole's hypothesis. Auk 97:38–49.

Ricklefs, R. E., M. Konrzewski & S. Daan. 1996. The relationship between basal metabolic rate and daily energy expenditure in birds and mammals. American Naturalist 147:1047–1071.

Ricklefs, R. E., D. D. Roby & J. B. Williams. 1986. Daily energy expenditure by adult Leach's storm-petrels during the nesting cycle. Physiological Zoology 59:649–660.

Root, T. 1988a. Environmental factors associated with avian distributional boundaries. Journal of Biogeography 15:489–505.

———. 1988b. Energy constraints on avian distributions and abundances. Ecology 69:330–339.

———. 1989. Energy constraints on avian distributions: A reply to Castro. Ecology 70:1183–1185.

Rosenzweig, M. L. 1968. The strategy of body size in mammalian carnivores. American Naturalist 80:299–315.

Ross, J. P. 1980. Seasonal variation of thermoregulation in the Florida pocket gopher, *Geomys pinetis*. Comparative Biochemistry and Physiology A 66:119–125.

Roth, V. L. 1990. Insular dwarf elephants: A case study in body mass estimation and ecological inference. In Body Size in Mammalian Paleontology, edited by J. Damuth & B. J. McFadden, 151–179. Cambridge University Press, Cambridge.

Ruben, J. A. 1995. The evolution of endothermy in mammals and birds: From physiology to fossils. Annual Review of Physiology 57:69–95.

Ruben, J. A., W. J. Hillenius, N. R. Geist, A. Leitch, T. P. Jones, P. J. Currie, N. R. Horner & G. Espe III. 1996. The metabolic status of some late Cretaceous dinosaurs. Science 273:1204–1207.

Ruben, J. A., T. D. Jones, N. R. Geist & W. J. H. Hillenius. 1997. Lung structure and ventilation in theropod dinosaurs and early birds. Science 278:1267–1270.

Rubner, M. 1883. Ueber den Einfluss der Köpergrosse auf Stoff- und Kraft wechsel. Zeitschrift für Biologie 19:535–562.

Sand, H., G. Cederlund & K. Danell. 1995. Geographical and latitudinal variation in growth patterns and adult body size of Swedish moose (*Alces alces*). Oecologia 102:433–442.

Sander, P. M. & M. Clauss. 2008. Sauropod gigantism. Science 322:200–201.

Sarrus, F. & J. F. Rameaux. 1839. Application des sciences accessoires et principalement des mathématiques à la physiologie générale. Bulletin de l'Académie Royale Médecine, Paris 3:1094–1106.

Savage, V. M., J. F. Gillooly, W. H. Woodruff, G. B. West, A. P. Allen, B. J. Enquist & J. H. Brown. 2004. The predominance of quarter-power scaling in biology. Functional Ecology 18:257–282.

Scheck, S. H. 1982. A comparison of thermoregulation and evaporative water loss in the hispid cotton rat, *Sigmodon hispidus texianus*, from northern Kansas and south-central Texas. Ecology 63:361–369.

Schleucher, E. 2004. Torpor in birds: Taxonomy, energetics, and ecology. Physiological and Biochemical Zoology 77:942–949.

Schmidt-Nielsen, K. 1970. Energy metabolism, body size, and problems of scaling. Federation Proceedings 29:1524–1532.

———. 1998. The Camel's Nose: Memoirs of a Curious Scientist. Island Press/Shearwater Books, Washington, DC.

Schmitz, O. J. & D. M. Lavigne. 1984. Intrinsic rate of increase, body size, and specific metabolic rate in marine mammals. Oecologia 62:305–309.

Scholander, P. F. 1940. Experimental investigations on the respiratory function in diving mammals and birds. Hvalrådets Skrifter 22:1–131.

———. 1990. Enjoying a Life in Science. University of Alaska Press, Fairbanks.

Scholander, P. F., R. Hock, V. Walters, F. Johnson & L. Irving. 1950a. Heat regulation in some arctic and tropical mammals and birds. Biological Bulletin 99:237–258.

Scholander, P. F., R. Hock, V. Walters & L. Irving. 1950b. Adaptation to cold in arctic and tropical mammals and birds in relation to body temperature, insulation, and basal metabolic rate. Biological Bulletin 99:259–271.

Scholander, P. F., V. Walters, R. Hock & L. Irving. 1950c. Body insulation of some arctic and tropical mammals and birds. Biological Bulletin 99:225–271.

Schuchmann, K.-L. 1999. Family Trochilidae (Hummingbirds). In Handbook of the Birds of the World, vol. 5, Barn-owls to Hummingbirds, edited by J. del Hoyo, A. Elliott & J. Sargatal, 468–680. Lynx Editions, Barcelona.

Scott, I., P. I. Mitchell & P. R. Evans. 1996. How does variation in body composition affect the basal metabolic rate of birds? Functional Ecology 10:307–313.

Seebacher, F., G. C. Grigg & L. A. Beard. 1999. Crocodiles as dinosaurs: Behavioural thermoregulation in very large ectotherms leads to high and stable body temperatures. Journal of Experimental Biology 202:77–86.

Seebacher, F., T. S. Schwartz & M. B. Thompson. 2006. Transition from ectothermy to endothermy: The development of metabolic capacity in a bird (*Gallus gallus*). Proceedings of the Royal Society of London B 273:565–570.

Shaw, W. T. 1925. Duration of the aestivation and hibernation of the Columbian ground squirrel (*Citellus columbianus*) and sex relation to the same. Ecology 6:75–81.

Shine, R. 2005. Life-history evolution of reptiles. Annual Review of Ecology, Evolution and Systematics 36:23–46.

Shkolnik, A. & A. Borut. 1969. Temperature and water relations in two species of spiny mice (*Acomys*). Journal of Mammalogy 50:245–255.

Shkolnik, A. & K. Schmidt-Nielsen. 1976. Temperature regulation in hedgehogs from temperate and desert environments. Physiological Zoology 49:56–64.

Sibley, C. & J. E. Ahlquist. 1990. Phylogeny and Classification of the Birds of the World: A Study in Molecular Evolution. Yale University Press, New Haven.

Sick, H. 1993. Birds in Brazil. Princeton University Press, Princeton, New Jersey.

Silver, H., N. F. Colovos, J. B. Holter & H. H. Hayes. 1969. Fasting metabolism of white-tailed deer. Journal of Wildlife Management 33:490–498.

Simms, D. A. 1979. North American weasels: Resource utilization and distribution. Canadian Journal of Zoology 57:504–520.

Simpson, G. G. 1980. Splendid Isolation: The Curious History of South American Mammals. Yale University Press, New Haven.

Skutch, A. F. 1949. Do tropical birds rear as many young as they can nourish? Ibis 91:430–458.

Slonin, A. D. 1952. Animal heat and its regulation in the mammalian organism. Academy of Sciences, USSR, Leningrad and Moscow.

Smith, R. L. & D. Rhodes. 1983. Body temperature of the salmon shark, *Lamna ditropis*. Journal of the Marine Biological Association, UK 63:243–244.

Smithers, R. H. N. 1971. The Mammals of Botswana. Memoirs of the National Museum of Rhodesia 4:1–340.

Sondaar, P. Y. 1977. Insularity and its effect on mammalian evolution. In Major Patterns in Vertebrate Evolution, edited by M. K. Hecht, P. C. Goody & B. M. Hecht, 671–707. Plenum, New York.

Soriano, P. J., A. Ruiz & A. Arends. 2002. Physiological responses to ambient temperature manipulation by three species of bats from Andean cloud forests. Journal of Mammalogy 83:445–457.

Sparti, A. 1990. Comparative temperature regulation of African and European shrews. Comparative Biochemistry and Physiology A 97:391–397.

Sparti, A. & M. Genoud. 1989. Basal rate of metabolism and temperature regulation in *Sorex coronatus* and *S. minutus* (Soricidae: Mammalia). Comparative Biochemistry and Physiology A 92:359–363.

Speakman, J. R. 2000. The cost of living: Field metabolic rates of small mammals. Advances in Ecological Research 30:177–297.

Speakman, J. R., R. M. McDevitt & K. R. Cole. 1993. Measurement of basal metabolic rates: Don't lose sight of reality in the quest for comparability. Physiological Zoology 66:1045–1049.

Speakman, J. R. & A. Rowland. 1999. Preparing for inactivity: How insectivorous bats deposit a fat store for hibernation. Proceedings of the Nutrition Society 58:123–131.

Stawski, C. & F. Geiser. 2010. Fat and fed: Frequent use of summer torpor in a subtropical bat. Naturwissenschaft 97:29–35.

Stawski, C., C. Turbill & F. Geiser. 2009. Hibernation by a free-ranging subtropical bat (*Nyctophilus bifax*). Journal of Comparative Physiology B 179:433–441.

Steadman, D. W. 2006. Extinction and Biogeography of Tropical Pacific Birds. University of Chicago Press, Chicago.

Steadman, D. W. & P. S. Martin. 2003. The Late Quaternary extinction and future resurrection of birds on Pacific islands. Earth-Science Reviews 61:133–147.

Stephenson, P. J. 1994a. Notes on the biology of the fossorial tenrec, *Oryzorictes hova* (Insectivora: Tenrecidae). Mammalia 58:312–315.

———. 1994b. Resting metabolic rate and body temperature in the aquatic tenrec *Limnogale mergulus* (Insectivora: Tenrecidae). Acta Theriologica 39:89–92.

Stephenson, P. J. & P. A. Racey. 1993a. Reproductive energetics of the Tenrecidae (Mammalia: Insectivora). I. The large-eared tenrec, *Geogale aurita*. Physiological Zoology 66:643–663.

———. 1993b. Reproductive energetics of the Tenrecidae (Mammalia: Insectivora). II. The shrew-tenrecs, *Microgale* spp. Physiological Zoology 66:664–685.

———. 1994. Seasonal variation in resting metabolic rate and body temperature of streaked tenrecs, *Hemicentetes nigriceps* and *H. semispinosus* (Insectivora: Tenrecidae). Journal of Zoology, London 232:285–294.

———. 1995. Resting metabolic rate and reproduction in the Insectivora. Comparative Biochemistry and Physiology A 112:215–223.

Stuenes, S. 1989. Taxonomy, habits, and relationships of the subfossil Madagascaran hippopotami *Hippopotamus lemerlei* and *H. madagascarensis*. Journal of Vertebrate Paleontology 9:241–268.

Swenson, J. E., M. Adamic, D. Huber & S. Stokke. 2007. Brown bear body mass and growth in northern and southern Europe. Oecologia 153:37–47.

Symonds, M. R. E. 1999. Life histories of the Insectivora: The role of physiology, metabolism, and sex differences. Journal of Zoology, London 241:315–337.

Taigen, T. L. 1983. Activity metabolism of anuran amphibians: Implications for the origin of endothermy. American Naturalist 121:94–109.

Tamarin, R. H. 1977. Demography of beach vole (*Microtus breweri*) and meadow vole (*Microtus pennsylvanicus*) in southeastern Massachusetts. Ecology 58:1310–1321.

Tannenbaum, M. G. & E. B. Pivorun. 1984. Differences in daily torpor patterns among three southeastern species of *Peromyscus*. Journal of Comparative Physiology B 154:233–236.

Taulman, J. F. & L. W. Robbins. 1996. Recent range expansion and distributional limits of the nine-banded armadillo (*Dasypus novemcinctus*) in the United States. Journal of Biogeography 23:635–648.

Taylor, C. R., N. C. Heglund & G. M. O. Maloiy. 1982. Energetics and mechanics of terrestrial locomotion. I. Metabolic energy consumption as a function of speed and body size in birds and mammals. Journal of Experimental Biology 97:1–21.

Taylor, C. R., K. Schmidt-Nielsen, R. Dmi'el & M. Fedak. 1971. Effect of hyperthermia on heat balance during running in the African hunting dog. American Journal of Physiology 220:823–827.

Taylor, C. R., K. Schmidt-Nielsen & J. L. Raab. 1970. Scaling of energetic cost of running to body size in mammals. American Journal of Physiology 219:1104–1107.

Tennyson, A. & P. Martinson. 2006. Extinct Birds of New Zealand. Te Papa Press, Wellington, New Zealand.

Thompson, S. D. 1985. Subspecific differences in metabolism, thermoregulation, and torpor in the western harvest mouse *Reithrodontomys megalotis*. Physiological Zoology 58:430–444.

———. 1987. Body size, duration of parental care, and the intrinsic rate of natural increase in eutherian and metatherian mammals. Oecologia 71:201–209.

———. 1988. Thermoregulation in the water opossum (*Chironectes minimus*): An exception that "proves" a rule. Physiological Zoology 61:450–460.

———. 1992. Gestation and lactation in small mammals: Basal metabolic rate and limits of energy use. In Mammalian Energetics: Interdisciplinary Views of Metabolism and Reproduction, edited by T. E. Tomasi & T. H. Horton, 213–259. Comstock, Ithaca, NY.

Tidemann, C. R., M. J. Vardon, R. A. Loughland & P. J. Brocklehurst. 1999. Dry season camps of flying-foxes (*Pteropus* spp.) in Kakadu World Heritage Area, north Australia. Journal of Zoology, London 247:155–163.

Tieleman, B. I., T. H. Dijkstra, J. R. Lasky, R. A. Mauck, G. H. Visser & J. B. Williams. 2006. Physiological and behavioural correlates of life-history variation: A comparison between tropical and temperate zone house wrens. Functional Ecology 20:491–499.

Tieleman, B. I. & J. B. Williams. 2000. The adjustment of avian metabolic rates and water fluxes to desert environments. Physiological and Biochemical Zoology 73:461–479.

Tieleman, B. I., J. B. Williams & P. Bloomer. 2002. Adaptation of metabolism and evaporative water loss along an aridity gradient. Proceedings of the Royal Society of London B 270:207–214.

Tófoli, C. F., F. Rohe & E. Z. Setz. 2009. Jaguarundi (*Puma yagouaroundi*) (Geoffroy, 1803) (Carnivora, Felidae) food habits in a mosaic of Atlantic rain forest and eucalypt plantations of southeastern Brazil. Brazilian Journal of Biology 69:871–877.

Tøien, Ø., J. Blake, D. M. Edgar, D. A. Grahn, H. C. Heller & B. M. Barnes. 2011. Hibernation in black bears: Independence of metabolic suppression of body temperature. Science 331:906–909.

Tomlinson, S., P. C. Withers & C. Cooper. 2007. Hypothermia versus torpor in response to cold stress in the native Australian mouse *Pseudomys hermannsburgensis* and the introduced house mouse *Mus musculus*. Comparative Biochemistry and Physiology A 148:645–650.

Tracy, C. R. 1972. Newton's law: Its application for expressing heat losses from homeotherms. BioScience 22:656–659.

Trevelyan, R., P. H. Harvey & M. D. Pagel. 1990. Metabolic rates and life histories in birds. Functional Ecology 4:135–141.

Trost, C. H. 1972. Adaptations of horned larks (*Eremophila alpestris*) to hot environments. Auk 89:506–527.

Tryjanowski, P., T. H. Sparks & P. Profus. 2005. Uphill shifts in the distribution of the white stork *Ciconia ciconia* in southern Poland: The importance of nest quality. Diversity and Distributions 11:219–223.

Tucker, V. A. 1965. Oxygen consumption, thermal conductance, and torpor in the California pocket mouse *Perognathus californicus*. Journal of Cellular and Comparative Physiology 65:393–404.

———. 1966. Diurnal torpor and its relation to food consumption and weight changes in the California pocket mouse *Perognathus californicus*. Ecology 47:245–252.

Turbill, C. & F. Geiser. 2008. Hibernation by tree-roosting bats. Journal of Comparative Physiology B 178:597–605.

Twente, J. H. & J. A. Twente. 1965. Effects of core temperature upon duration of hibernation of *Citellus lateralis*. Journal of Applied Physiology 20:411–416.

Tyndale-Biscoe, C. H. 1973. Life of Marsupials. American Elsevier, New York.

Van Mierop, L. H. S. & S. M. Barnard. 1978. Further observations on thermoregulation in the brooding female *Python molurus bivittatus* (Serpentes: Boidae). Copeia 1978:615–621.

Van Valen, L. 1973. Pattern and the balance of nature. Evolutionary Theory 1:31–49.

Vardon, M. J., P. S. Brocklehurst, J. C. Z. Wolnarski, R. B. Cunningham, C. F. Donnelly & C. R. Tidemann. 2001. Seasonal habitat use by flying-foxes, *Pteropus alecto* and *P. scapulatus* (Megachiroptera), in monsoonal Australia. Journal of Zoology, London 253:523–535.

Vartanyan, S. I., V. E. Garutt & A. V. Sher. 1993. Holocene dwarf mammoths from Wrangel Island in the Siberian Arctic. Nature 362:337–340.

Vogt, F. D. & R. Lynch. 1982. Influence of ambient temperature, nest availability, huddling, and daily torpor on energy expenditure in the white-footed mouse *Peromyscus leucopus*. Physiological Zoology 55:56–63.

von Hoesslin, H. 1888. Ueber die Ursache der Scheinbaren Abhängigkert des Umsatzes von der Grösse der Körperoberfläche. DuBois-Reymond Archive für Anatomie und Physiologie 11:323–379.

Walker, J. M., A. Garber, R. J. Berger & H. C. Heller. 1979. Sleep and estivation (shallow torpor): Continuous processes of energy conservation. Science 204:1098–1100.

Wallace, A. R. 1876. The Geographical Distribution of Animals. Harper and Brothers, New York.

Wang, L. C. H. 1979. Time patterns and metabolic rates of natural torpor in the Richardson's ground squirrel. Canadian Journal of Zoology 57:149–155.

Wang, Z., T. P. O'Connor, T. Heshka & S. B. Heymsfield. 2001. The reconstruction of Kleiber's law at the organ-tissue level. Journal of Nutrition 131:2967–2970.

Wasser, J. S. 1986. The relationship of energetics of falconiform birds to body mass and climate. Condor 88:57–62.

Wasserman, D. & D. J. Nash. 1979. Variation in body size, hair length, and hair density in the deer mouse *Peromyscus maniculatus* along an altitudinal gradient. Holarctic Ecology 2:115–118.

Watts, P. D. & C. Cuyler. 1988. Metabolism of the black bear under simulated denning conditions. Acta Physiologica Scandinavica 134:149–152.

Watts, P. D. & S. E. Hansen. 1987. Cyclic starvation as a reproductive strategy in the polar bear. Symposium of the Zoological Society of London 57:305–318.

Watts, P. D. & C. Jonkel. 1988. Energetic cost of winter dormancy in grizzly bear. Journal of Wildlife Management 52:654–656.

Watts, P. D., N. A. Øritsland & R. J. Hurst. 1987. Standard metabolic rate of polar bears under simulated denning conditions. Physiological Zoology 60:687–691.

Weathers, W. W. 1979. Climatic adaptation in avian standard metabolic rate. Oecologia 42:81–89.

Weaver, J. C. 1983. The improbable endotherms: The energetics of the sauropod dinosaur *Brachiosaurus*. Paleobiology 9:173–182.

Webb, P. I. & J. D. Skinner. 1995. Resting metabolism and thermal conductance in southern Africa's smallest rodent, the pygmy mouse (*Mus minutoides*). Zeitschrift für Säugetierkunde 60:251–254.

Webb, S. D. 1976. Mammalian faunal dynamics of the great American interchange. Paleobiology 2:220–234.

———. 1978. History of savanna vertebrates in New World. 2. South America and Great Interchange. Annual Review of Ecology and Systematics 9:393–426.

———. 1985a. Late Cenozoic mammal dispersals between the Americas. In The Great American Biotic Interchange, edited by F. G. Stehli & S. D. Webb, 357–386. Plenum Press, New York.

———. 1985b. The interrelationships of tree sloths and ground sloths. In The Evolution and Ecology of Armadillos, Sloths, and Vermilinguas, edited by G. G. Montgomery, 105–112. Smithsonian Institution Press, Washington, DC.

———. 1991. Ecogeography and the Great American Interchange. Paleobiology 17:266–280.

Weigold, H. 1979. Köpertermperatur, Sauerstoffverbrauch und Herzfrequenz bei *Tupaia belangeri* Wagner, 1841 im Tagesverlauf. Zeitschrift für Säugetierkunde 44:343–353.

Weiner, J. 1977. Energy metabolism of the roe deer. Acta Theriologica 22:3–24.

———. 1989. Metabolic constraints to mammalian energy budgets. Acta Theriologica 34:3–35.

———. 1992. Physiological limits to sustainable energy budgets in birds and mammals. Trends in Ecology and Evolution 7:384–388.

Weisman, A. 2007. The World Without Us. St. Martin's Press, New York.

West, G. B., J. H. Brown & B. J. Enquist. 1997. A general model for the origin of allometric scaling laws in biology. Science 276:122–126.

———. 1999. The fourth dimension of life: Fractal geometry and allometric scaling of organisms. Science 284:1677–1679.

Wetterer, A. L., M. V. Rockman & N. B. Simmons. 2000. Phylogeny of phyllostomid bats (Mammalia: Chiroptera): Data from diverse morphological systems, sex chromosomes, and restriction sites. Bulletin of the American Museum of Natural History 248:1–200.

White, C. R., T. M. Blackburn, G. R. Martin & P. J. Butler. 2007a. Basal metabolic rate of birds is associated with habitat temperature and precipitation, not primary productivity. Proceedings of the Royal Society of London B 274:287–293.

White, C. R., P. Cassey & T. M. Blackburn. 2007b. Allometric exponents do not support a universal metabolic allometry. Ecology 88:315–323.

White, C. R., T. M. Blackburn & R. S. Seymour. 2009. Phylogenetically informed analysis of the allometry of mammalian basal rate of metabolism supports neither geometric nor quarter-power scaling. Evolution 63:2658–2667.

White, C. R. & R. S. Seymour. 2003. Mammalian basal metabolic rate is proportional to body mass$^2_3$. Proceedings of the National Academy of Sciences, USA 100:4046–4049.

———. 2004. Does basal metabolic rate contain a useful signal? Mammalian BMR allometry and correlations with a selection of physiological, ecological, and life-history variables. Physiological and Biochemical Zoology 77:929–941.

White, T. A. & J. B. Searle. 2007. Factors explaining increased body size in common shrews (*Sorex araneus*) on Scottish islands. Journal of Biogeography 34:356–363.

Whittow, G. C. & E. Gould. 1976. Body temperature and oxygen consumption of the pentail tree shrew (*Ptilocercus lowii*). Journal of Mammalogy 57:754–756.

Whittow, G. C., E. Gould & D. Rand. 1977. Body temperature, oxygen consumption, and evaporative water loss in a primitive insectivore, the moon rat, *Echinosorex gymnurus.* Journal of Mammalogy 58:233–235.

Wiersma, P., A. Muñoz-Garcia, A. Walker & J. B. Williams. 2007. Tropical birds have a slow pace of life. Proceedings of the National Academy of Sciences, USA 104:9340–9345.

Wikelski, M., L. Spinney, W. Schelsky, A. Scheuerlein & E. Gwinner. 2003a. Slow pace of life in tropical sedentary birds: A common-garden experiment on four stonechat populations from different latitudes. Proceedings of the Royal Society of London B 270:2383–2388.

Wikelski, M., E. M. Tarlow, A. Ralm, R. H. Diehl, R. P. Larkin & G. H. Visser. 2003b. Costs of migration in free-flying songbirds. Nature 423:704.

Williams, J. B. 1987. Field metabolism and food consumption of savannah sparrows during the breeding season. Auk 104:277–289.

Williams, J. B., A. Muñoz-Garcia, S. Ostrowski & B. I. Tieleman. 2004. A phylogenetic analysis of basal metabolism, total evaporative water loss, and life-history among foxes from desert and mesic regions. Journal of Comparative Physiology B 174:29–39.

Williams, J. B. & B. I. Tieleman. 2005. Physiological adaptation in desert birds. BioScience 55:416–425.

Williams, J. B., P. C. Withers, S. D. Bradshaw & K. A. Nagy. 1991. Metabolism and water flux of captive and free-living Australian parrots. Australian Journal of Zoology 39:131–142.

Williams, T. M., J. Haun, R. W. Davis, L. A. Fulman & S. Kohin. 2001. A killer appetite: Metabolic consequences of carnivory in marine mammals. Comparative Biochemistry and Physiology A 129:785–796.

Williams, T. M. & G. A. J. Worthy. 2002. Anatomy and physiology: The challenge of aquatic living. In Marine Mammal Biology, edited by A. R. Hoelzel, 73–97. Blackwell Science, Oxford.

Willis, C. K., J. D. Skinner & H. G. Robertson. 1992. Abundance of ants and termites in the False Karoo and their importance with the diet of the aardvark *Orycteropus afer.* African Journal of Ecology 30:322–334.

Wilson, E. O. 2002. The Future of Life. Alfred A. Knopf, New York.

Wimsatt, W. A. 1969. Transient behavior, nocturnal activity patterns, and feeding efficiency of vampire bats (*Desmodus rotundus*) under natural conditions. Journal of Mammalogy 50:233–244.

Winter, Y. & O. von Helversen. 1998. The energy cost of flight: Do small bats fly more cheaply than birds? Journal of Comparative Physiology B 168:105–111.

Withers, P. C., K. C. Richardson & R. D. Wooller. 1990. Metabolic physiology of euthermic and torpid honey possums, *Tarsipes rostratus.* Australian Journal of Zoology 37:685–693.

Withers, P. C. & J. B. Williams. 1990. Metabolic and respiratory physiology of an arid-adapted Australian bird, the spinifex pigeon. Condor 92:961–969.

Wolverton, S., M. A. Huston, J. H. Kennedy, K. Cagle & J. D. Cornelius. 2009. Conformation to Bergmann's rule in white-tailed deer can be explained by food availability. American Midland Naturalist 162:403–417.

Worthy, T. H., A. J. Anderson & R. E. Molnar. 1999. Megafaunal expression in a land without mammals: The first fossil faunas from terrestrial deposits in Fiji. Senckenbergiana Biologica 79:237–242.

Yáñez, J. L., J. C. Cárdenas, P. Gezelle & F. M. Jaksic. 1986. Food habits of the southernmost mountain lions (*Felis concolor*) in South America: Natural versus livestocked ranges. Journal of Mammalogy 67:604–606.

Yom-Tov, Y. 1987. The reproductive rate of Australian passerines. Australian Wildlife Research 14:319–330.

———. Clutch size of passerines at mid-latitudes: The possible effect of competition with migrants. Ibis 136:161–165.

Zink, R. M. & J. V. Remsen, Jr. 1986. Evolutionary processes and patterns of geographic variation in birds. In Current Ornithology, vol. 4, edited by R. F. Johnston, 1–69. Plenum Press, New York.

TAXONOMIC INDEX

SUBJECT INDEX